# EQUILIBRIUM
## A Chemistry of Solutions

# EQUILIBRIUM
## A Chemistry of Solutions

Thomas R. Blackburn

*Hobart and William Smith Colleges*

**HOLT, RINEHART AND WINSTON, INC.**

*New York   Chicago   San Francisco   Atlanta   Dallas*
*Montreal   Toronto   London   Sydney*

Library of Congress Catalog Card number 69-20458
SBN: 03-069785-9

Printed in the United States of America

1    2    3    4    5    6    7    8    9

# Preface

A variety of human enterprises — medicine, chemistry, geology, biology, oceanography, physiology, and all the fringes and combinations of these fields — have a common need to understand the chemical behavior of solutions. This book is intended to serve as an introduction to solution equilibrium for students whose interests lie in one or more of the above directions.

It has become increasingly accepted that analytical chemistry is not a necessary context for understanding solution equilibrium. On the one hand, many of those whose work will involve equilibrium (for example, pre-medical students) are only tangentially interested in the tactics and lore of analysis; on the other, the eternal expansion of non-classical methods of analysis leaves little room in a modern analytical course for the leisurely contemplation of, say, ionic solubility, which is an important topic in geochemistry. It is the author's hope that equilibrium as presented here will appear relevant to the concerns of students majoring in many areas of science, and for a longer time than the one-term course for which this text was designed.

I have addressed this book, as I suppose all authors do, to an imagined reader who may or may not be realistic. Objectively, he has had at least one course in chemistry, including the meanings of chemical symbols and re-actions as well as an introduction to modern structural theories. He has also taken, but perhaps forgotten, enough high school mathematics to understand algebra, logarithms, and graphs. He has not had an introduction to thermodynamics as such, though if he has, he may be interested in the thermodynamic background to equilibrium summarized in Appendix 2. Subjectively, I assume that the reader is as indifferent to the formal language of science as I; and that he will collaborate in his learning by working through the many exercises which are scattered through the body of the text. Some of these exercises are quite simple, others are difficult, and some present developments of ideas which are not otherwise treated in the text. All of them are designed to pull the reader into discourse with the author.

I have tried to arrange the major topics in an order which will allow efficient development of later material in terms of earlier. It is for this reason that Chapter 2, on Brønsted acid-base equilibria, appears disproportionately long: it develops the fundamental algebraic and graphical tools as well as describing proton-transfer equilibria, while succeeding chapters use both the logical and chemical concepts of Chapter 2 to streamline their discourse.

In Chapters 2, 3, and 4 I have interwoven some simple structurally-based rationales of reactivity with the purely mathematical consequences, in order to approach justice to the wholeness of the idea of chemical equilibrium. In doing so I have done less than justice, no doubt, to modern quantum mechanics and to those enigmatic systems, the molecules themselves.

The discussions of chromatographic techniques and of the experimental details of electrochemical measurements have been kept to a bare minimum, not through any desire to minimize their importance and their intellectual challenge, but because I have felt it best to keep the scope of this book limited to the fundamental concepts of equilibrium. I hope that a sketch of the manifold applications will serve the reader as a token of the riches which await him in further study.

I would like to express my thanks to Professors Donald West, Robert Hatala, Dennis Evans and Henry A. Bent, who have read the entire manuscript and made many helpful suggestions and criticisms; to Mr. Roger Purcell, Geological Survey Department, Indiana University, who drafted the figures; to Wellesley College for a small grant toward the expense of preparing the manuscript; to my teachers Richard W. Ramette and James J. Lingane, whose erudition and patience far exceed my own, and whose influence on this book they, and most other chemists will recognize; to my wife, Roberta Dayton Blackburn, who consulted on matters of English style while keeping the household quiet and the book growing; and to my students at Carleton, Wellesley, and Hobart and William Smith colleges, who have continuously and in good spirit corrected my crude over-and underestimations of their powers and refined my understanding of equilibrium with their insights.

THOMAS R. BLACKBURN

# Contents

# 3    Coordination Equilibria    *71*

# 4    Solubility Equilibria    *93*

# 5    Distribution Equilibrium across a Liquid Phase Boundary    *113*

# 6  Oxidation-Reduction Equilibria and Electrochemical Cells    *141*

## Appendices    *176*

## Index    *215*

*Acknowledgments:*


p. 1: Quotation from Gerard Manley Hopkins,
"*Inversnaid.*" *Poems of Gerald Manley Hopkins,*
Oxford University Press.


p. 113: Quotation from Alistair McLean's
*Night Without End.* Doubleday & Company, Inc.


Data in Tables A3.1-A3.3 selected from
The Chemical Society's *Stability Quotients,*
by Lars Gunnar Sillen and Arthur E. Martell.


Data in Table A3.4 from Wendell M. Latimer,
*Oxidation Potentials,* 2nd ed., Englewood Cliffs,
N.J.: Prentice-Hall, Inc., 1952; and Ernest
H. Swift, *Introductory Quantitative Analysis,*
Englewood Cliffs, N.J.: Prentice-Hall, Inc., 1950.

# EQUILIBRIUM
A Chemistry of Solutions

# 1 Order out of Chaos

What would the world be, once bereft
Of wet and of wildness? Let them be left,
O let them be left, wildness and wet;
Long live the weeds and the wilderness yet.

Gerard Manley Hopkins, *"Inversnaid"*

## Order, Chaos, and Equilibrium

It has been said that a chemist is a scientist who loves molecules. It is fortunate for chemists, then, that molecules are so incredibly numerous. The Nobel laureate who synthesized chlorophyll created something like $10^{21}$ replicas of the object of his devotion. Microscopic samples of matter are composed of unimaginably large numbers of inconceivably tiny objects, each of which may exist in a large number of states of energy, configuration, position, and motion. Like many more familiar aggregations of independent objects, a chunk of matter is an extremely chaotic situation.

If we drop a box of marbles, commission an "action" finger painting from a 3-year-old artist, or call a meeting of a 100-member committee, we are not surprised when the results have little of what we conventionally call *order*. An ordered array of a large set of independent objects is only one of many possible arrangements of the objects. For example, consider alphabetical seating in classrooms. Such an orderly arrangement has about one chance in 4 million of occurring in a 10-member class, when the order of last names is produced by such random processes as friendship, visual handicaps, etc. Even if a group is alphabetized, they will quickly become disordered if allowed to wander of their own accord. An observer knowing nothing of the workings of probability might conclude that there is among groups of people a "drive" toward disorder in alphabetical arrangements. There is, however, no particular disordering *force;* there are simply many more disordered arrangements than there are ordered ones. If all arrangements are allowed to occur with equal probability, the chances are overwhelming that the one we find will be nonalphabetical.

Consideration of the alphabetizing of various numbers of named objects will convince you that the more chaotic a situation is, the more ways there are of realizing it. For example, in our hypothetical 10-member class, the likelihood of an alphabetical arrangement with only one member out of order is 10 times as great as that of a completely correct arrangement, but it is still only one in about 400,000. And, of course, the likelihood that some two members will be in alphabetical order is a virtual certainty; in fact, the chances that no two members will sit in alphabetical order is again one in 4 million. (Why?)

In the molecular world, with its vast numbers of molecules, each with large numbers of possible states, the possibilities for disorder are colossal. The result is that the only processes ever observed are those which are a change from much disorder to even more disorder, because this is equivalent to saying that the only processes we are *likely* to observe are those involving a change from a highly probable state to an even more highly probable state. The "likely" in the previous sentence is italicized because of the size of the likelihood involved; in a visibly large sample of matter, the most likely state is so overwhelmingly more probable that states requiring even a trivial increase in microscopic order have virtually zero probability.

It is no denial of this principle of ever-increasing microscopic disorder that we often observe processes in which *part* of the world becomes more orderly. In admiring the beauty of spontaneously created frost crystals on a windowpane, we do not see the increased disorder of the thermal motion acquired by molecules in the surrounding air when the water which froze gave up its heat of crystallization. Taken as a whole, the universe became more disorderly on the molecular level when that frost formed. Thus, it often happens that a given process achieves disorder in one place at the cost of creating order in another.

To see how the related ideas of randomness, probability, and disorder affect chemical reactions, let us consider an example which, though simple, includes the essential features of much more complex reactions.

Formic acid,

$$H-C\underset{OH}{\overset{O}{<}}$$

when present as a solute in a nonpolar solvent or in the gas phase, dimerizes through interactions called hydrogen bonds between the electron-rich oxygen atom of one molecule with the electron-poor hydrogen of another:

$$2H-C\overset{\displaystyle O}{\underset{\displaystyle OH}{}} \longrightarrow H-C\overset{\displaystyle O\cdots H-O}{\underset{\displaystyle O-H\cdots O}{}}C-H \qquad (1\text{-}1)$$

The dotted lines in the structure of the dimer indicate the hydrogen bonds.

We may now ask whether a given collection of formic acid molecules will exist as single molecules or as dimers. There are certainly arguments on both sides, and in fact a compromise is reached. If we look at the system from the standpoint of randomness, we see that a collection of independent formic acid molecules has more opportunity for randomness (and thus a higher probability) than the same number of molecules arranged in pairs; each formic acid molecule that is a member of a dimer has lost its randomness in position and velocity with respect to its partner. On the other hand, we may argue that the forces between formic acid molecules, e.g., the coulombic force between the somewhat negative oxygen and the somewhat positive hydrogen, should bring them together and hold them in a dimer. Why should these forces not cause our collection of molecules to exist exclusively as dimers?

Let us digress for a moment and consider a more familiar example of a force causing a change of position. If we hold a brick above the floor, the gravitational field gives it potential energy with respect to the floor. If we release the brick, it reaches the floor promptly, with no second thoughts about disorder. However, if we replace the brick with a tennis ball, we find that it does not reach a motionless state as rapidly as the brick did. It bounces, and it may take several seconds to "obey" the force of gravity by coming to rest on the floor. One of those super-bouncy rubber balls would take even longer, and a perfectly bouncy ball would take infinite time to settle onto a (perfectly hard) floor. That is, gravitational force has no power to effect a *permanent* change in the position of a perfectly bouncy ball.

Let it be clear that we do not claim that the perfectly bouncy ball bounces forever because by doing so it achieves randomness of position (although it certainly does). The point is that forces *alone* are powerless to produce a permanent change in a system. Only when the potential energy of the bouncing ball is converted into something other than kinetic energy of that same ball is rest on the floor possible. Eventually, of course, the potential and kinetic energy of real objects is converted into heat in their more or less inelastic collisions with the floor, and having neither potential nor kinetic energy relative to the floor, they come to rest.

The heat energy in the resting ball and the floor is prevented from being reconverted into kinetic energy and starting the whole process over again because this energy is so *randomly* distributed in the disorderly thermal vibrations of the molecules of the floor and ball. The probability of the heat spontaneously flowing through the floor and the ball back to the site of a bounce, and there converting itself to a "palpable hit," which could set the ball bouncing again, is very small. It would require that all the motions of the incredible number of molecules affected by the bounce be exactly reversed *by chance*, so that the spreading wave of energy which radiated from the site of the bounce would reconverge to that site and form a tiny depression, into which would crouch a suitably deformed ball, and to which would converge also the sound wave ("plop!") which originally spread out from the site of the bounce. Then with a "!polp" the ball and depression in the floor would straighten out and, aided by the absorbed sound energy, the ball would fly into the air.

The above scenario for a reversed motion picture is what we never observe. The universe, like a no-nonsense movie, runs in one direction only: the direction of increased randomness, because this is the direction of increased probability. This increase of randomness makes permanent the changes initiated by potential energy.

Returning to our formic acid dimerization, we see that when reaction 1-1 goes to the right as written, one sort of randomness (free motion of any formic acid molecule with respect to any other) is exchanged for another (the random translations, rotations, and vibrations of thermal energy). A large collection of molecules will reach a compromise of maximum randomness in which some of the molecules remain monomers, some are present as dimers (and the resulting energy is distributed at random through the collection and its surroundings as heat), and (most important) monomers and dimers are mixed together randomly.

When a state of maximum possible disorder is reached, no more probable state is accessible to the system, and thus no further net change can occur. The system (in our example, the mixture of monomers and dimers) is in a state of *chemical equilibrium*. This is not to say that no further changes occur on the molecular level. A particular formic acid molecule will find itself now a monomer, now a member of a dimer, as the forward and reverse motions of reaction 1-1 continue. But in a state of chemical equilibrium, the *rates* of the forward and reverse reactions are equal, so that there is no net change in the relative amounts of products and reactants.

It was first observed empirically, and then shown to follow from more general observations called the laws of thermodynamics (see Appendix 2), that at equilibrium a restraint exists on the relative amounts of products and reactants. For our example of the formic acid dimerization, it takes the form

$$\frac{[\text{dimer}]}{[\text{monomer}]^2} = \text{constant} \qquad (1\text{-}2)$$

where the brackets indicate concentrations in moles per liter. For a general reaction,

$$a\text{A} + b\text{B} + \cdots = c\text{C} + d\text{D} + \cdots \qquad (1\text{-}3)$$

$$\frac{[C]^c[D]^d \dots}{[A]^a[B]^b \dots} = \text{constant} \qquad (1\text{-}4)$$

The equals ($=$) sign in Eq. 1-3 indicates that the reaction has reached equilibrium. Since this is by no means an instantaneous process in many reactions, this is an important condition. Equation 1-2 and its generalized analog are only true under the condition of equilibrium.

**Deviations from ideal behavior.** Having come this far, we must now face the fact that Eqs. 1-2 and 1-4 are only approximately true. This is so because 1-1 and 1-3 do not show all of the interactions the reactants and products undergo. For example, in a reaction involving ions, long-range coulombic forces between ions will have an effect on the tendency of a given species to react, and the value of the equilibrium "constant" will depend on the kind and number of all ions present in the solution, including those which may not take part in the reaction. The usual way out of this dilemma is to postulate the existence of a function called the *activity* of a species (for example, of A in Eq. 1-3), which is related to its concentration via the equation

$$(A) = [A] \cdot f_A \qquad (1\text{-}5)$$

Parentheses around a chemical symbol indicate the activity of that species. The quantity $f_A$ is simply the factor by which the activity differs from the concentration. The activity is ultimately defined through the relationship between the number of moles of component A in the solution and the disorder of the universe. (Those who have studied the thermodynamic functions are referred to Appendix 2, where this relationship is discussed.) The factor $f_A$, called the activity *coefficient*, is defined by Eq. 1-5.

A function of the form of the left-hand side of Eq. 1-2 or 1-4 is called the *proper quotient of concentrations;* when it is measured under the condition of equilibrium, it is called the equilibrium quotient, and given the symbol $Q$. The proper quotient of *activities* taken at equilibrium is called the equilibrium *constant*, and given the symbol $K$:

$$K = \frac{(C)^c(D)^d \dots}{(A)^a(B)^b \dots} \qquad (1\text{-}6)$$

Unlike $Q$, $K$ is truly constant under all conditions of solution composition at any particular temperature. This follows from the manner in which the activities are defined. Frankly, activities were defined in the first place so that equations of the simple form of (1-6) and others like it would be true.

Since $K$ is independent of solution conditions (though it is dependent on temperature), it is valid for any experiment and is worth recording and tabulating in the chemical literature. For any particular experiment, with its own unique solution composition, the value of $Q$ for some reaction will be more or less different from $K$. Approximate methods for calculating the value of $Q$ from a knowledge of $K$ and of the composition (mainly the number and type of the ions present in it) are outlined in Appendix 2. Depending on the charge of the reactants and products in a reaction, and on the strength of their interactions with each other and with other nonreacting species in the solution, $Q$ may differ from $K$ for that reaction only negligibly, or by a large factor. For example, $Q$ and $K$ for reaction 1-1 are not likely to differ greatly under any circumstances, because neither the reactants nor the products are charged. Weak van der Waals interactions arising from the polarity of the formic acid molecule might cause $Q$ to vary from $K$ by a few per cent. On the other hand, in ionic reactions carried out in the presence of many other ions, $Q$ and $K$ may differ typically by a factor of 2 or 3, and in extreme cases by as much as a factor of 10 or more. Because of this situation, and because only approximate methods for relating $Q$ and $K$ are known, both tabulated $K$ and calculated $Q$ should be regarded as more or less accurate guesses at the truth.

Since $Q$ relates "real," or at least intuitively palpable quantities, molarities, it is usually of greater experimental interest than $K$, even though the latter is more closely related to fundamental thermodynamic quantities. In this book, we will concentrate on the relationship between equilibrium molarities and equilibrium quotients $Q$, assuming that the latter are known either through calculation from a tabulated $K$, or by direct experimental measurement, which is possible in many cases. In my opinion, these relationships contain ample food for thought, and express more truth (and thus beauty) than the minor problem of the relationship between $Q$ and $K$ for a particular reaction.

## Concentration and its Units

Before simply accepting Eq. 1-4 uncritically, you should wonder why we have used concentrations, which express quantity of reactant per unit volume, rather than simply quantity, for example in units of moles or grams, or perhaps the volume, in units of liters, cubic centimeters, or bushels. The answer is that Eq. 1-3 expresses an equilibrium with respect to a transformation of matter from one state to another, and that the driving force for such transformations is generally proportional not to the total number of molecules of matter involved, but to the degree to which they are *confined* in each state; that is, the number of molecules or of moles per unit volume. For example, compare the force with which air seeks to escape from a balloon before and after decreasing its volume by squeezing it. Clearly, it is not the quantity of air in the balloon but that quantity divided by the volume of the balloon that is proportional to the tendency of the air to leave it.

If, then, a condition of equilibrium reflects equal and opposite tendencies for products to become reactants, and for reactants to revert to products, as well as equal rates for those opposing processes, we should expect that concentrations should be used in the equilibrium constant, and not simply the total quantities of reactants and products involved.

Concentrations are one example of an *intensive* variable, that is, a variable whose value is independent of the total quantity of material present, as opposed to *extensive* variables which are proportional to the total quantity. Other examples of intensive variables are pressure, temperature, and density; examples of extensive variables are mass, volume, and weight. In general, values of intensive variables determine whether a given process can occur. Heat flows only across a difference in temperature, and fluids move only through a difference in pressure. The use of the intensive concentration variable in an equilibrium constant reflects the fact that substances react only when there is a difference in a quantity (called "chemical potential") which is related to the concentrations of the substances involved.

Units of concentration can be assembled in a variety of ways, some of which make use of a distinction (fundamentally arbitrary) between the substance present in major proportion (called the solvent) and those present in minor proportions (called solutes). When we dissolve salt in a colleague's coffee, we have no hesitation in identifying, if called on, the salt as solute and the water in the coffee as solvent. However, consider the vodka martini. The composition of this mixture is essentially 50% ethanol and 50% water, with traces of flavoring substances. Choice of one of the major components as solvent and the other as a solute is clearly arbitrary.

The following are the commonly used units of concentration, all of them used to describe the concentration of a particular solute:

**Molarity: moles of solute per liter of solution.**    Molarities are most commonly used when solutions are handled with volume-measuring instruments such as pipets, burets, or volumetric flasks. Because the volume of a solution changes when its temperature changes, molarities have the disadvantage of not being independent of temperature. However, volumetric glassware is convenient and simple to use, and molarity is the most frequently encountered unit.

**Formality: formula weights per liter of solution.**    Formality differs from molarity only in that the concept of formula weight is substituted for the number of moles. This insures intellectual honesty in cases in which the number of moles of the substance whose molecular formula is given is in doubt or different from the number of formula weights of that substance in the solution. For example, if we dissolve one formula weight of NaCl in sufficient water to make one liter of solution, it is speaking loosely to say that the concentration of NaCl in that solution is 1 $M$, since the ionic dissociation of the salt reduces the concentration of the species NaCl to a very small number. However, one may say that the concentration of NaCl in that solution is 1 formal (abbreviated 1 $F$) without making any commitment as to the fate of the NaCl once dissolved. Being a volume-based concentration unit, formality shares a disadvantage of molarity in being temperature-dependent.

**Molality: moles of solute per kilogram of solvent.** Molality is used widely in physical chemistry, partly because it is independent of temperature, and partly because it is more simply related to chemical potentials than are the volume-based concentration units. The convenience of use with volumetric glassware is sacrificed.

**Mole fraction: moles of solute divided by the total number of moles of all components.** Mole fraction is especially useful when, as in the case of the vodka martini, there is no particular basis for calling one component the solvent. It shares the advantages and disadvantages of molality as discussed above.

**Normality: equivalents of solute per liter of solution.** Normality is based on the idea of the "equivalent," that is, the number of moles of some substance which will react with one mole of the solute in question. Since the nature of the reacting substance is not always clear, and rarely specified, normality is probably the largest single source of confusion, lost time, and irritation in a laboratory which uses it. For example, a 1 $M$ solution of $HC_2O_4^-$ is 1 normal with respect to reaction with a base and 2 normal with respect to reaction with an oxidizing agent. If one desires to synthesize the tris-oxalato complex of some transition metal ion [for example $Cr(C_2O_4)_3^{3-}$] it would be $\frac{1}{3}$ normal. The proper use of normality can save calculation time in an analysis (although in my experience, the multiplicative factor is as often used inverted as right-side up), by virtue of concealing one step of the stoichiometric calculation in the labelling of the solution. Normalities have no place in fundamental chemical studies.

## Calculations with the Equilibrium Quotient

Let us illustrate the usefulness of equilibrium quotients first by using one to calculate the equilibrium composition of a solution of formic acid monomers and dimers, and second to illustrate how that composition may vary with the total (formal) concentration of formic acid present in that solution.

Suppose that we have a solution containing $C$ formula weights of formic acid per liter; that is, the concentration of formic acid is $C\ F$. For every mole of monomer M in solution, there is one formula weight of acid present; for every mole of dimer D, there are two formula weights present. That is,

$$C = [M] + 2[D] \qquad (1\text{-}7)$$

Also, the relative quantities of monomer and dimer must obey the equation

$$Q = \frac{[D]}{[M]^2} \qquad (1\text{-}8)$$

If we solve Eq. 1-8 for [D] and substitute the result in Eq. 1-7, we obtain

$$C = [M] + 2Q[M]^2 \qquad (1\text{-}9)$$

Equation 1-9 is a quadratic equation in a single unknown [M]. One way to solve it (although it is not the only one) is to use the quadratic formula. For this case, it has the form:

$$[M] = \frac{-1 \pm (1 + 8QC)^{1/2}}{4Q} \qquad (1\text{-}10)$$

EXERCISE 1.1:   Fill in the missing algebra to obtain Eqs. 1-9 and 1-10.

Since the quantity within the parentheses in the numerator is positive, the solutions of Eq. 1-10 will be non-imaginary (they should be because Eq. 1-9 follows from a real physical problem!). Further, because both $Q$ and $C$ are by their nature positive numbers, the square root will be a number greater than 1, which means that there will be one negative and one positive root for any values of $Q$ and $C$. Being uninterested in the concept of negative molarity, we choose the positive root.

Let us make things a little more concrete with a real set of numbers:

EXERCISE 1.2:   In benzene as solvent at 25°C, the dimerization equilibrium quotient $Q$ for formic acid is approximately 120. Calculate the [M] and, from Eq. 1-7, [D] for a 0.01 $F$ solution of formic acid. (Answer: [M] $= 4.7 \times 10^{-3}\ M$; [D] $= 2.6 \times 10^{-3}\ M$.) Do the same calculation for 0.1 $F$ formic acid. How do you interpret this result as compared to that for $C = 0.01\ F$?

## Shifting Equilibria and Le Chatelier's Principle

There is a principle, derivable from the postulates of thermodynamics or from common sense, that states that when a system at equilibrium is subjected to a stress of any sort, the system will change in such a direction as to relieve the stress. To put *Le Chatelier's principle* in simpler terms, things go where you push them. For example, suppose you have a cylinder with a tightly fitting piston, containing a liquid in equilibrium with its vapor. If the piston is moved inward, decreasing the volume available to the vapor, the pressure increases and exceeds the equilibrium vapor pressure. Some vapor condenses, decreasing the number of moles of vapor, thus decreasing the pressure in the vapor phase until equilibrium is re-established. By Le Chatelier's principle, the system has shifted in the direction which relieved the stress of increased pressure on the system.

EXERCISE 1.3:   What would have happened if the system had responded in the other direction?

To take another example, suppose that you plunge the cylinder momentarily into an ice bath, allowing some heat to flow out through the walls from the system inside. The temperature of the liquid-vapor system drops, and it is again subjected to a stress and is out of equilibrium, since the vapor pressure of any liquid decreases when the temperature decreases. The system responds by condensing some of the vapor. Whenever a vapor condenses, heat (the heat of vaporization) is liberated, as van der Waals forces translate potential energy into random thermal vibrations. The temperature of the liquid-vapor system now rises until equilibrium is reestablished.

EXERCISE 1.4:   This time what would have happened if the system had shifted in the opposite direction? Le Chatelier's principle is an example of *negative feedback*. How many other examples can you think of?

In a chemical equilibrium, Le Chatelier's principle implies that the composition of an equilibrium mixture will change when the mixture is disturbed by any sort of stress, including the addition or removal of one of the reactants or products of the equilibrium. For a start, consider the general equilibrium

$$aA + bB = cC + dD$$

whose equilibrium quotient is

$$Q = \frac{[C]^c[D]^d}{[A]^a[B]^b}$$

If we add some additional quantity of one component, say C, to an equilibrium mixture of A, B, C, and D, the proper quotient of concentrations will momentarily be different from (in this case, larger than) $Q$. The system will adjust to this stress by removing some of the added C; i.e., the reaction will go to the left as written, causing the concentrations of C and D to decrease as the concentrations of A and B increase. Both changes cause the proper quotient of concentrations to decrease until it again equals Q. The process is more vividly described by saying that the addition of C caused the equilibrium to shift to the left. We will find this idea and terminology very useful in qualitative discussions of equilibria, so be sure that you understand what lies behind it.

EXERCISE 1.5: Suppose a few millimoles of D are removed from the equilibrium mixture of A, B, C, and D. Describe the response of the system both in the laborious language of proper quotients of concentrations, and as a "shift" of the equilibrium.

To return to our concrete example, let us see the effect of Le Chatelier's principle on the monomer-dimer equilibrium of formic acid. Those who worked exercise 1.2 will recall that the fraction of the formic acid present as monomer was smaller at $C = 0.1$ $F$ than at $C = 0.01$ $F$. This behavior is another manifestation of Le Chatelier's principle. When the formal concentration of formic acid is increased, the stress on the system may be thought of as "crowding" in the solution; the system's response is to alleviate the crowding by converting some monomers into dimers. Doubtless some of you have had to share facilities which you would just as soon have had to yourself. This is a socio-political manifestation of Le Chatelier's principle.

The behavior of the monomer-dimer equilibrium over a range of values of $C$ is graphed in Fig. 1-1a. Note that the trend established in Exercise 1.2 continues smoothly over the range of concentrations graphed. The more formic acid present in the solution, the smaller is the fraction of it present as monomer.

Because the interesting part of Fig. 1-1a tends to be in the left-hand side of the graph, we can change the abscissa from $C$ to log $C$ (see Appendix 1 if you are not familiar with logarithms). This has the effect of spreading out the region below $C = 0.01$, and compressing the relatively dull region above $C = 0.03$. It does not change the qualitative message of the graph, which is the important feature.

## The Temperature Dependence of Equilibrium Constants

We remarked in passing that the numerical values of equilibrium constants depend on the temperature. The precise relationship between $K$ and $T$ may be derived from those generalizations on energy and disorder, the laws of thermodynamics, but we have already noticed an important quali-

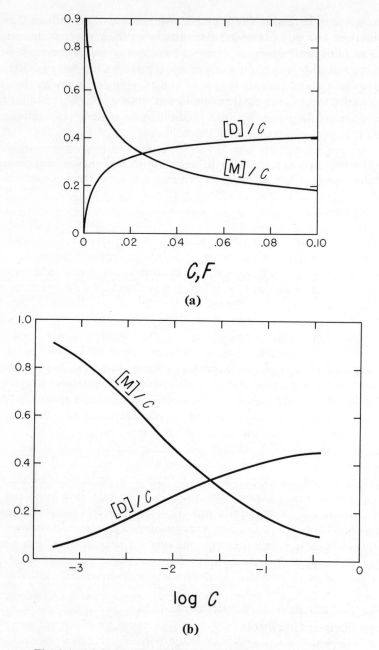

**Fig. 1-1. (a)** Fractions of formic acid in benzene solution present as monomer ($[M]/C$) and as dimer ($[D]/C$) as a function of total formal concentration $C$. **(b)** Fractions present as a function of log $C$. Note that the low-concentration part of the graph is more legible on a log $C$ axis.

tative guide in Le Chatelier's principle. Suppose we begin with a reaction at equilibrium at a given temperature, and place it in a warmer environment. Heat will flow into the equilibrium system; how can the system respond in order to relieve this stress? Clearly, the system can allow the equilibrium to shift in such a direction that some of the incoming heat is absorbed. This can be achieved by altering the compromise between disorder and potential energy toward the side of disorder, allowing part of the incoming energy to appear as increased internal potential energy. In the example of the monomer-dimer equilibrium of formic acid, $K$ decreases with increasing temperature, as more of the system is converted to high-potential energy, high-freedom monomer molecules.

Although it is not always as easy as in the case of formic acid to see at a glance which side of a reaction represents higher disorder, it is always true that when the temperature of a chemical system is raised, any equilibrium in that system will shift in such a direction as to absorb heat; and *vice versa* for a decrease in temperature. For most reactions occurring in solution, the effect of temperature on the equilibrium constant is appreciable, but not huge; a change of a few per cent per degree is typical.

## Water Solutions of Ions

Despite its apparently all-embracing title, and the generality of the discussion so far, this book will be concerned mainly with the properties of solutions of ionic substances in water. Of the three states of matter, liquids are the least well understood. Pure liquids are much more straightforward than solutions, and aqueous solutions of ions are among the most complex solutions. However, a surprisingly good understanding of these complex systems can be achieved by using the relatively simple ideas of chemical equilibrium. Because ionic solutions constitute not only the larger portion of the surface of the earth, but also most of the systems with which chemists and biologists deal (including the chemists and biologists themselves), an elementary understanding of them seems worthwhile.

When it was first advanced, the idea that many substances on dissolving in water dissociate into charged particles was met by considerable opposition. Our present tendency to laugh at the conservatism of this reception may result from a failure to appreciate the magnitude of interionic forces. The complete separation of a mole of sodium chloride into two side-by-side piles of sodium ions and chloride ions at a distance of 4 inches could only be maintained by exerting a force of about $10^{14}$ tons; even if one of the piles were removed to the surface of the moon, it would still attract the other with a force of over half a ton. To give a more germane example, the average separation of a sodium ion and a chloride ion in a 1 $F$ solution of NaCl is roughly $10^{-7}$ cm. At this distance, in a vacuum, the force between the two ions would be about $2 \times 10^{-5}$ dyne. This force (about one hundred-thousandth the weight of a mosquito) is small by macroscopic standards; but operating on the two ions, it would cause them to come together at an acceleration of about $5 \times 10^{17}$ cm/sec$^2$, which is about $5 \times 10^{14}$ times as rapidly as a brick accelerates when you drop it on your toe.

What effects, then, can possibly operate against these powerful forces to allow the existence of such important ionic solutions as seawater, battery acid, and blood? There are three main considerations: disorder, the dielectric constant of water, and an energy of interaction between solvent molecules and the solute ions. Of these, the last is the most important, but it will be instructive to consider each.

Disorder, it should be apparent, may increase within the solution when an orderly crystalline solid is converted to a collection of separate particles, free to move in random directions at random velocities throughout the much larger volume of the solution. However, what counts is not the disorder of the solution alone, but the disorder of the entire universe; that is, the disorder of the solution plus its surroundings. Large forces must be overcome in separating ions, and if the energy required to do this were to be drawn from the surroundings, the latter would cool down so much in supplying the energy that their increased order at a lower temperature would more than cancel out the disorder achieved by the solution. Only when little or no energy needs to be drawn from the surroundings to form a solution is the disorder of the solution the controlling factor. An example of this is the fact that iodine — a nonionic, nonpolar solid held together only by weak interactions between $I_2$ molecules — readily dissolves in such nonpolar solvents as $CCl_4$ and benzene.

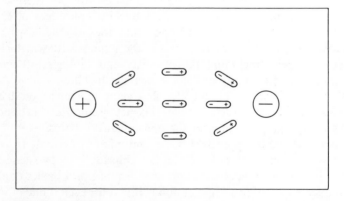

**Fig. 1-2.** Dipolar molecules of a dielectric solvent aligned between oppositely charged solute ions. Thermal vibrations will disrupt this idealized arrangement more or less depending on the temperature and the association of the solvent.

The calculations of the force between oppositely charged ions which we discussed above were made on the assumption that there was no material substance between the ions; i.e., these are forces *in vacuo*. If, in fact, the space between the ions is filled with a substance whose molecules are polar (as are the molecules of water), the forces are smaller by a factor called the *dielectric constant* of the substance. Figure 1-2 depicts the effect of an electric field on the ordering of solvent molecules between two ions dissolved in a polar solvent. The net effect of all of the aligned dipoles between the two ions is an electric field oppositely directed to that of the two ions, and partly cancelling it. The forces between ions in water solutions are only about one 80th as large as they would be in a vacuum because of this effect.

This large value — 80 — for the dielectric constant of water deserves a closer look because, although it is a very high value, certainly the highest value of any common liquid, water molecules are exceeded in polarity by those of many substances with smaller dielectric constants. For example, a molecule of chloromethane has approximately the same dipole moment as one of water, but the dielectric constant of liquid water is over six times as large as that of chloromethane. The explanation for this discrepancy is that random thermal vibrations tend to disrupt the orderly alignment shown in Figure 1-2, and only a fraction of any solvent's potential dielectric constant is actually observed. In water there is a special force aiding the alignment of the molecules: hydrogen bonding between the slightly positive hydrogens of one molecule and lone electron pairs on the oxygens of a neighboring molecule. This interaction, which is not as strong as an ordinary covalent bond, but is stronger than typical dipole-dipole interactions, allows the idealized alignment of Figure 1-2 to be approached more nearly for water and other associated liquids than for polar, but nonassociated, liquids like chloromethane, nitromethane, and chloroform.

Even if we divide the above interionic forces by 80, we are still left with large numbers. One other factor accounts for the existence of ionic solutions despite the large forces between ions: solvation energy. When a dipole (such as a polar-solvent molecule) encounters an ion, the end of the dipole similar in charge to the ion is repelled, and the other end is attracted (Fig. 1-3). The end being repelled is farther from the ion than the end being attracted, and since electric forces vary inversely with the square of the distance, a net attractive force exists between the ion and the dipole. The decrease in potential energy accomplished by this attractive force when the ion and dipole come together in large measure compensates for the energy required to separate the ions in the solution.

The extent to which the energy of solvation compensates for the energy required to separate ions may be seen in two examples: When solid potassium nitrate is dissolved in water, the solution becomes cool; that is, some energy must be supplied from the surroundings in order for the crystal to be broken up at a given temperature because the energy released by hydration

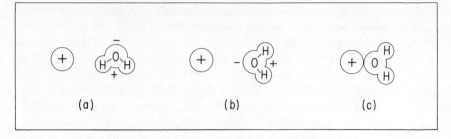

(a)            (b)            (c)

Fig. 1-3. (a) Positive ion and dipolar water molecule. (b) Orientation of the water molecule in the ion's electrical field. (c) Attraction of the water molecule to the ion.

of the ions is not quite sufficient to compensate completely for the energy required to separate the potassium and nitrate ions. On the other hand, quite a large rise in temperature occurs when sodium hydroxide dissolves in water, because the solvent-solute interaction releases more than enough energy to break up the ionic solid, and the excess appears as heat.

## Suggestions for Further Reading

This book assumes that the reader has had a modern introductory course in chemistry, including elementary molecular structure and reactions. A number of texts which center on the theme of structure and reactivity are now available. Particularly recommended is Milton K. Snyder, *Chemistry: Structure and Reactions.* New York: Holt, Rinehart & Winston, 1966.

Many of the suggested readings for this and other chapters are intended for advanced readers. I feel that the student who has understood this book will be prepared to tackle these more difficult works:

Bent, H. A., *The Second Law, An Introduction to Classical and Statistical Thermodynamics.* New York: Oxford University Press, 1965. A book written with an eye to fresh and arresting developments of the time-honored truths of thermodynamics. Chapters 1–14 are especially relevant to our discussion of order and chaos.

DeFord, D. D., "The Reliability of Calculations Based Upon the Law of Chemical Equilibrium." *J. Chem. Ed., 31*, 460 (1954). Good, short discussions of the various pitfalls, from unsuspected side-reactions to activity corrections, which limit the reliability of equilibrium calculations. A very healthy antidote to a sickly fascination for super-accurate calculations based on erroneous assumptions and/or data.

Hunt, J. P., *Metal Ions in Aqueous Solution.* New York: W. A. Benjamin Co., 1963. The title describes it. The book expands the introduction to the subject in this chapter. Some knowledge of thermodynamics is assumed.

Lewis, G. N., and M. Randall, *Thermodynamics*, Second Ed., revised by K. S. Pitzer and L. Brewer. New York: McGraw-Hill Publishing Co., 1961. The book has the distinction of having the most-quoted preface in textbook history, and is a central book in every chemist's library. It is intended for advanced students of chemistry, but Chapter 8 on entropy and probability can be read by anyone, and will supplement our treatment while giving the novice the incomparable flavor of this book.

# 2 Proton Transfer Equilibria

## Introduction

The most ubiquitous and, by force of numbers, the most important class of reactions taking place in water solutions are those in which a hydrogen ion is transferred from one molecule or ion to another. Since hydrogen atoms, with one electron, are the simplest atoms, a positive hydrogen ion is the same as a hydrogen nucleus. In the predominant fraction of hydrogen atoms, which is the isotope of mass one, the nucleus is a single proton, and hydrogen-ion transfer is generally known as proton transfer. After centuries of progressively refined definitions of what is meant by the terms acid and base, Brønsted proposed that reactions involving the transfer of a hydrogen ion be called acid-base reactions; the proton donor is called a Brønsted acid, and the proton acceptor a Brønsted base. (The terms "donor" and "acceptor" put a polite face on the transaction; the proton goes where it can bond most strongly, and the Brønsted base has more realistically been termed a proton snatcher, and the acid a proton loser.) When a Brønsted acid loses a proton, it is converted into a species which can gain one; that is, every acid has its *conjugate base*, a substance which differs from the acid only in that a proton is missing. Conversely, of course, every Brønsted base has its *conjugate acid*. For example, in the acid-base reaction

$$HF + NH_3 = F^- + NH_4^+ \qquad (2\text{-}1)$$

the HF is converted to its conjugate base $F^-$, and the $NH_3$ to its conjugate acid $NH_4^+$.

## The Autoprotolysis of Water

In the preceding chapter we have seen that water molecules strongly interact through the formation of hydrogen bonds between a somewhat positive hydrogen atom of one molecule and the electron-rich oxygen atom of a second (Figure 2-1). Such an interaction is a natural preliminary to the

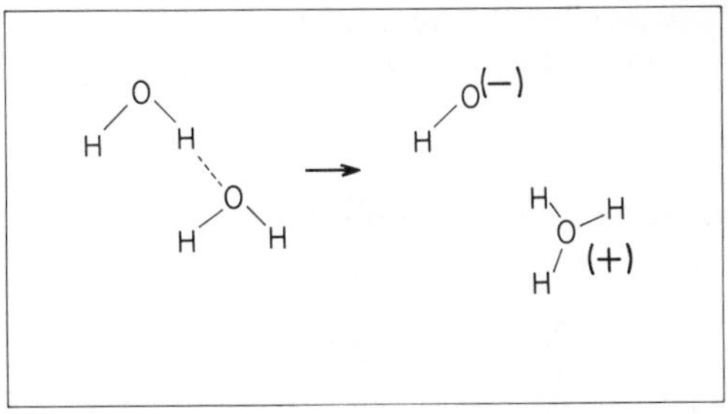

**Fig. 2-1.** Transfer of a proton between two hydrogen-bonded water molecules. The signs in parentheses represent partial ionic charges responsible for the polarity of water molecules.

outright transfer of a proton from the first water molecule to the second, where it forms a covalent bond:

$$2H_2O = H_3O^+ + OH^- \qquad (2\text{-}2)$$

The species $H_3O^+$, which is the conjugate acid of water, is called the hydronium ion. There is some uncertainty about the number of water molecules involved in the hydration of a proton; some results indicate that it is four, giving the conjugate acid of water the formula $H_9O_4^+$ and the structure shown in Figure 2-2. For our purposes the symbol $H_3O^+$, or simply $H^+$, will be used to indicate the solvated proton. The structure in Figure 2-2 is not intended to represent an unchanging entity. Waters of hydration will depart to be replaced by others, and clearly a proton can transfer easily from the central oxygen to a peripheral one by moving along a hydrogen bond as shown in Figure 2-1, thus shifting the center of the structure. Indeed, this sort of leap-frogging of the identity of the hydronium ion is invoked to account for the fact that protons appear to be unusually fast moving in water, having mobilities 5 to 10 times those of ordinary positive ions.

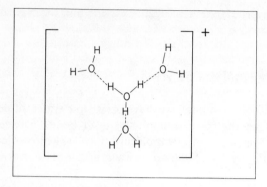

**Fig. 2-2.** A possible structure for $H_9O_4^+$. The solid lines are covalent bonds, and the dotted lines are strong hydrogen bonds.

**The pH of pure water.**    The equilibrium quotient for the autoprotolysis of water (Eq. 2-2) is

$$Q = \frac{[H_3O^+][OH^-]}{[H_2O]^2} \qquad (2\text{-}3)$$

However, it is a convention in physical chemistry that when the solvent participates in a chemical reaction in solution, its concentration in the equilibrium quotient is expressed in units of mole fraction. This convention exists because the properties of the solvent in a dilute solution are practically the same as those of the pure solvent, because its mole fraction is close to unity. The departure of the chemical properties of the solvent in a solution from those of the pure solvent is closely related to its mole fraction in the solution. Another advantage of the convention is that because the mole fraction of the solvent is very close to unity, it can, as a reasonable approximation, be left out of products and quotients, such as the equilibrium quotient in Eq. 2-3. With this convention, then, Eq. 2-3 may be rewritten as

$$Q_w = [H_3O^+][OH^-] \qquad (2\text{-}4)$$

where the subscript $w$ indicates that this particular equilibrium quotient is for the autoprotolysis of water. At 25°C, $K_w$ has the value $1.0 \times 10^{-14}$.

If we note that, according to Eq. 2-2, one hydronium ion is formed for every hydroxide ion, we are able to calculate the concentrations of $H_3O^+$ and $OH^-$ formed by the autoprotolysis of pure water. To be mathematically formal, we have two unknowns ($[H_3O^+]$ and $[OH^-]$) and two equations, 2-4 and the condition

$$[H_3O^+] = [OH^-] \qquad (2\text{-}5)$$

which follows from Eq. 2-2. Combining these, we arrive at

$$[H_3O^+]^2 = Q_w = 1.0 \times 10^{-14} \qquad (2\text{-}6)$$

where we have ignored salt effects on $Q_w$ at the very low ionic strength of pure water. We will specify that we are not concerned with the negative root to Eq. 2-6, and, with that, take leave of mathematical rigor. The physically significant solution to 2.6 is, of course,

$$[H_3O^+] = 1.0 \times 10^{-7} M \qquad (2\text{-}7)$$

We convert Eq. 2-7 to pH by taking the negative logarithm:

$$pH = 7.0$$

that is, the pH produced by the autoprotolysis of pure water is 7.0. (The reader who is unfamilar with logarithms is referred to Appendix 1.)

## The Effect of Temperature and Ionic Strength on the pH of Pure Water

The numerical value 7.0 for the pH of pure water is a result of the numerical value of $Q_w$, $10^{-14}$. We noted in Chapter 1 that equilibrium quotients in general have values dependent on both temperature and the composition of the solution, and that equilibrium constants (such as $K_w$) are dependent on temperature. Because this is our first equilibrium calculation, we should consider these two perturbing effects; what we say in this case will apply as well to all subsequent discussions

Using Le Chatelier's principle as a guide to the effect of temperature on equilibrium constants, we concluded in Chapter 1 that increasing the equilibrium temperature of a reaction will shift that equilibrium (and thus change $K$) in the direction such that heat is absorbed by the reaction. If you have ever observed the effect of mixing a strong acid with a strong base, you will recall that a considerable quantity of heat is evolved by the reaction

$$H_3O^+ + OH^- = 2H_2O$$

which is the reverse of the autoprotolysis reaction. We may conclude that the autoprotolysis of water proceeds with the *absorption* of heat, and that raising the temperature should therefore increase the degree of autoprotolysis and the value of $K_w$. This expectation is correct: $pK_w$ is 14.94 at 0°C, 13.9965 at 25°C, 13.53 at 40°C, and 13.02 at 60°C. You should calculate, on the basis of the foregoing discussion, the pH of pure water at each of these temperatures.

The value of $Q$ for an ionic reaction depends on both the corresponding $K$ and the presence of other ions in the solution. This effect, mentioned in Chapter

1, is discussed at length in Appendix 2. For the present case, we note that in Eq. 2-2 all ions are on the right-hand side of the reaction. The electrical effects of other ions will be manifested as an increase or a decrease in the stability of the products of autoprotolysis, $H_3O^+$ and $OH^-$. As long as the solution is not too crowded with ions, the presence of other ions (for example, from an ionic solute) tends to stabilize the separation of charge which occurs when Eq. 2-2 goes to the right. Negative ions in solution tend to gather around the $H_3O^+$, and positive ions around the $OH^-$, obeying the well-known proclivities of opposite charges. When $H_3O^+$ and $OH^-$ are surrounded by "atmospheres" of oppositely charged ions, their energies and thus their reactivities are decreased. The rightward direction of reaction 2-2 is not much affected, but the leftward direction is hindered, and the $Q_w$ is increased. From Eq. A2-17*,[1] Appendix 2, one may calculate, for example, that $Q_w$ is larger than $K_w$ by a factor of 1.66 in a 0.1 $F$ solution of a 1-1 salt like KCl. In such a solution, then, $pQ_w$ would be less than $pK_w$ by log 1.66, or 0.22. The $pQ_w = 13.78$, and pH = 6.89 in a neutral solution. Because Eq. A2-17* is an approximate equation, both $pQ_w$ and pH are uncertain in the second decimal place.

There are several points to note about the solution to this simple problem. One is that we have chosen to solve for $[H_3O^+]$; we could equally well have chosen to eliminate this variable and arrived at the equally correct solution $[OH^-] = 1.0 \times 10^{-7}$. There is an informal tradition that the acid-base status of a solution is given by the concentration of protonated, rather than of deprotonated, solvent molecules even though in an autoprotolysis equilibrium the two concentrations are related by Eq. 2-4, and one value is as informative as the other. Furthermore, Eq. 2-5 embodies a set of ideas which we will find useful later in solving more complex problems. We arrived at Eq. 2-5 by invoking stoichiometry; that is, since Eq. 2-2 is the only source of either $H_3O^+$ or $OH^-$ in the solution, the concentrations of these two must be present in the proportions in which the species are produced in Eq. 2-2, i.e., one to one.

We could equally well have arrived at Eq. 2-5 by invoking the idea that the solution as a whole must be electrically neutral, since the enormous electrostatic forces described in Chapter 1 prevent the removal of more than a negligible amount of charge from the solution. Then the sum of the positive charges in solution must equal the sum of the negative charges. There is only one positive species and one negative species in this solution, and Eq. 2-5 follows. The condition of equal total positive and negative charges in a solution is known as electroneutrality.

Finally, Eq. 2-5 may be derived from a more specific idea, that of proton balance. Beginning with a solution containing only neutral water molecules as a reference level, we look for species representing either an excess of protons or a deficiency of protons with respect to the reference level

[1]Here and throughout the book, starred equation numbers indicate approximate or idealized relationships.

species. Clearly, there is in this problem just one species representing proton excesses, $H_3O^+$, and one representing proton deficiencies, $OH^-$. We may equate excesses and deficiencies, because any proton which occurs on an excess species must have left a deficient species. Again, Eq. 2-5 follows simply. We will presently see the use of proton balance in more complex problems.

## Solutions of Brønsted Acids and Bases

In the preceding sections we have seen that water is capable of acting as an acid and as a base. When other basic or acidic substances are dissolved in water, we may expect them to react with the solvent to a greater or lesser extent. Let us now consider the composition of solutions produced by the reaction of acids with water. Remaining noncommittal about the nature of the species, except for its possession of an ionizable hydrogen, we consider the acid HX, where X is any atom or group of atoms:

$$HX + H_2O = X^- + H_3O^+ \tag{2-8}$$

represents its reaction with water. The equilibrium quotient for this reaction is, if we remember to assume unit mole fraction for the solvent,

$$Q_a = \frac{[H_3O^+][X^-]}{[HX]} \tag{2-9}$$

We may now ask what pH is produced by the presence of this acid in the solution. Since the answer depends not only on the quantity of the acid present, but also on the value of the characteristic quotient $Q_a$ for HX, we will briefly discuss structural factors in HX which influence its $K_a$. (Since reaction 2-8 is an interaction of two species, we should expect that the $K_a$ also depends on the solvent. We will return to this matter in a later section of this chapter.)

At this point, we introduce some ambiguous but prevalent nomenclature. Acids are categorized as *strong* or *weak*, depending on the value of the $K_a$. Since $K_a$ can have any of a virtual continuum of values, there is no sharp dividing line between the two categories. Generally, if the $K_a$ is so large that the concentration of the undissociated acid HX present at equilibrium is undetectably small, and the acid appears to be completely dissociated at any concentration, we call the acid strong. On the other hand, if only a very small fraction of the acid is dissociated at equilibrium, the acid is clearly weak. Where you care to draw the line between strong and weak is your own choice, but an acid with a $K_a$ of greater than about $10^{-4}$ is considered a rather strong weak acid, and one with a $K_a$ of less than roughly $10^{-8}$ is considered a very weak acid. A "typical" weak acid might have a $K_a$ between $10^{-4}$ and $10^{-10}$; acetic acid, with a $K_a$ of about $2 \times 10^{-5}$, is most often used for elementary discussions of weak acids. Note that a certain illogic lurks in this nomenclature: If the molecule binds its proton strongly, so that it dissociates only slightly, it is called weak, but if it can barely hold on to it, it is called strong.

**Factors influencing acid strength.** The ease with which a proton leaves a given species depends primarily on the atom to which it is bonded, and secondarily on more remote influences on that atom. Hydrogens bonded to electronegative elements are removed relatively easily as positive ions, since part of the work of producing the ion is done by the polarization of the H-X bond. The acidity of hydrides in a given row of the periodic chart increases going to the right along the row. For example, all of the binary compounds of hydrogen with the elements of Groups VI and VII are Brønsted acids. Perhaps surprisingly, within a given group the acidity of the hydrides increases as one goes down the group, i.e., as electronegativity decreases. This anomaly results because in comparing, for example, HF with HI, which is a much stronger acid, we are comparing not only the polarities of the two compounds, but also the length of the hydrogen-halogen bonds. Because the outer electronic shells of the elements get larger as their principal quantum number increases, the radius of the elements in a particular group generally increases as one goes down the group. Since an iodine atom is larger than a fluorine atom, the H-I bond is longer than the H-F bond. It is a general fact that a long bond is a weak one, and the proton-iodide bond is easy to break not because of the relatively low electronegativity of iodine, but in spite of it.

By far the largest class of acids is that in which the ionizable hydrogen is attached to oxygen. (In fact, during the late 18th century it was felt that the presence of oxygen in a compound was a necessary condition for acidity. The name *oxygen* means acid-former.) The acidic strength of a compound containing an —OH group depends strongly on the other atoms attached to the oxygen, ranging from compounds such as the alcohols (or the hydroxide ion!), which are hardly acidic, to sulfuric and perchloric acids, which are among the strongest acids known. The most important factors appear to be those listed in Table 2.1.

*Electronegativity.* The effect of the element attached to the oxygen that carries the ionizable hydrogen drawing electrons to itself is transmitted, either through bonds or through space, to the O-H bond, polarizing it so as to make departure of a proton easy. The electronegativity of a particular atom is strongly dependent on its oxidation state, so that, for example, in the series $HClO$, $HClO_2$, $HClO_3$, $HClO_4$, acid strength increases uniformly

**TABLE 2.1** MOLECULAR STRUCTURE AND STRENGTHS OF OXYGEN ACIDS[a]

*I. Effect of Electronegativity of Atom Adjacent to Oxygen*

| ACID | STRUCTURE | $pK_a$ at 25°C |
|---|---|---|
| Hypoiodous | H—O<br>   \<br>    I | 10.6 |
| Hypobromous | H—O<br>   \<br>    Br | 8.62 |
| Hypochlorous | H—O<br>   \<br>    Cl | 7.30 |
| Chlorous | H—O<br>   \<br>    Cl—O | 1.96 |
| Chloric | H—O<br>   \<br>    Cl—O<br>     \|<br>     O | −1 (?) |
| Perchloric | H—O   O<br>   \  /<br>    Cl<br>   /  \<br>  O    O | (Very strong) |

*II. Effect of Remote Electron-Withdrawing Groups*

| ACID | STRUCTURE | $pK_a$ at 25°C |
|---|---|---|
| Acetic | $H_3CC$ (with =O above and OH below) | 4.76 |
| Iodoacetic | $ICH_2C$ (with =O above and OH below) | 3.12 |
| Bromoacetic | $BrCH_2C$ (with =O above and OH below) | 2.90 |

*II. Effect of Remote Electron-Withdrawing Groups (Cont.)*

| ACID | STRUCTURE | $pK_a$ at 25°C |
|------|-----------|----------------|
| Chloroacetic | $ClCH_2C$ $\overset{O}{\underset{OH}{\diagdown}}$ | 2.87 |
| Dichloroacetic | $Cl_2CHC$ $\overset{O}{\underset{OH}{\diagdown}}$ | 1.3 |
| Trichloroacetic | $Cl_3CC$ $\overset{O}{\underset{OH}{\diagdown}}$ | 0.7 (?) |
| Butyric | $H_3C(CH_2)_2C$ $\overset{O}{\underset{OH}{\diagdown}}$ | 4.82 |
| 4-Chlorobutyric | $H_2C(CH_2)_2C$, Cl, OH $\overset{O}{\diagdown}$ | 4.52 |
| 3-Chlorobutyric | $H_3CCHCH_2C$, Cl, OH $\overset{O}{\diagdown}$ | 4.05 |
| 2-Chlorobutyric | $H_3CCH_2CHC$, Cl, OH $\overset{O}{\diagdown}$ | 2.86 |

[a]Some of the data in this table were obtained from noncritical compilations and should be considered rough values illustrative of the discussion in this chapter. For more critically assembled constants, the source listed in Appendix 3 is to be preferred.

with the oxidation state of the chlorine (in all of these, the hydrogen is attached to an oxygen atom). In a given oxidation state, acid strength increases with the element's inherent electronegativity. For example, the $K_a$'s for the following acids increase in the order: HOI < HOBr < HOCl.

*Remote electron-withdrawing groups.* The general effect of electronegative groups on the ease of ionic dissociation of an O-H bond is observable even when the electron-withdrawing group is relatively remote from the O-H bond; and again the more electronegative the group the larger, generally, is its effect on the acidity of the OH group. For example, acidity increases in the series: acetic acid < iodoacetic acid < bromoacetic acid < chloroacetic acid. Successive addition of further electronegative groups produces a smooth increase in acidity: chloroacetic acid < dichloroacetic acid < trichloroacetic acid. The more remote the electronegative atom from the ionizable hydrogen, the weaker is the effect: butyric acid < 4-chlorobutyric < 3-chlorobutyric < 2-chlorobutyric.

*Electronic stabilization of the conjugate base.* In many oxygen acids it can be argued that the effect of electron-withdrawing groups operates not so much through polarization of the O-H bond before ionization as in the stabilizing delocalization of the negative charge resulting from the departure of the proton. Thus although ordinary alcohols are negligibly acidic, phenols are more acidic because the negative charge on a phenoxide anion is shared among more atoms than that on say, an ethoxide anion. Similar arguments can be advanced for most oxygen acids. In the present state of bonding theory it is difficult to say whether the effect of an electronegative atom in an acid is primarily due to the weakening of the ionizable bond or to the stabilizing of the conjugate base. Both effects undoubtedly play a role.

**Hydrated metal ions as Brønsted acids.**    Metal ions in aqueous solutions, as we have seen, interact strongly with water molecules to form hydrated ions (Fig. 1-3, Chapter 1). We could represent a typical assemblage by $M(OH_2)_n^{+m}$. This species contains $2n$ oxygen-hydrogen bonds, which will be more or less polarized toward the metal ion, depending on the electrical field of the latter. The more polarized the O-H bond, the easier it is to break it to form a hydrogen ion. Hydrated metal ions, in short, are capable of behaving as Brønsted acids in solution. We should expect that their strength as acids would increase with increasing charge on the metal ion, and this is approximately true. (We will have more to say about the strength with which metal ions attract electron-rich species in Chapter 3.)

For the equilibrium

$$M(OH_2)_n^{+m} + H_2O = [M(OH_2)_{n-1}OH]^{+m-1} + H_3O^+$$

the following $pQ_a$'s have been observed: $Na^+$, 14.6; $Ag^+$, 11.7; $Mg^{2+}$, 11.4; $Ni^{2+}$, 10.6; $Cu^{2+}$, 7.5; $Hg^{2+}$, 2.49; $Fe^{3+}$, 2.17; $Al^{3+}$, 5.02; and $Tl^{3+}$, 1.14. Thus the hydrated aluminum ion is about as strong an acid as acetic acid, and the ferric ion is about as strong as phosphoric acid. Solutions of these ions are as acidic as if the acid had been a conventional Brønsted acid. (Some deodorant products cause rashes because they use aluminum salts as drying agents; vinegar is about as acidic, and is said not to draw flies.)

The reaction by which a hydrated metal ion acts as a Brønsted acid has traditionally been called *hydrolysis*. This name makes some chemists uncomfortable because hydrolysis ("cleavage by means of water") has a distinct meaning in connection with such reactions as

$$C_2H_5C \overset{\displaystyle O}{\underset{\displaystyle OCH_3}{\big\backslash}} + H_2O = C_2H_5C \overset{\displaystyle O}{\underset{\displaystyle OH}{\big\backslash}} + HOCH_3$$

in which the ester methyl propionate is *hydrolyzed* (cleaved) by water into two parts. No one has found a satisfactory term to substitute for hydrolysis in describing the acid behavior of hydrated metal ions. As a result, semantic purists use quotation marks to convey their discomfort: thus, "hydrolysis."

## pH in Solutions of Strong Acids

If the $K_a$ of an acid is quite large, reaction 2-8 lies so far to the right that the quantity of undissociated HX present at equilibrium is practically undetectable. This means that, for this type of acid, the $K_a$ is extremely difficult to measure experimentally, and large uncertainties attend any tabulated value. Fortunately, calculation of the $[H_3O^+]$ in a solution of a strong acid does not require a knowledge of the $K_a$. Since, by hypothesis, virtually no HX is present at equilibrium, one mole each of $X^-$ and $H_3O^+$ are formed for every formula weight of HX added to the solution. If the formal concentration of HX is $C$, then to an excellent approximation,

$$[X^-] = [H_3O^+] = C \qquad (2\text{-}10^*)$$

This equation is an approximation for two reasons: first, because no equilibrium can be driven absolutely to completion, a trace of HX must be present in solution, and second because HX is not the only source of $H_3O^+$. For a large class of acids (Table 2.2), the error of ignoring the slight incompleteness of dissociation is nil. Equation 2-10* is also approximate because it ignores the $H_3O^+$ produced by the autoprotolysis of water.

A proton balance equation starting from the reference level $HX + H_2O$ gives us the exact equation:

$$[H_3O^+] = [X^-] + [OH^-] \qquad (2\text{-}11)$$

Equation 2-11 results from equating total proton excesses to total proton deficiencies relative to the reference level. It may also be read, "one $H_3O^+$ is produced for each $X^-$ (reaction 2-8) and for each $OH^-$ (reaction 2-2)."

TABLE 2.2   SOME STRONG ACIDS

| ACID | STRUCTURE |
|------|-----------|
| Perchloric | $HOClO_3$ |
| Sulfuric[a] | $(HO)_2SO_2$ |
| Hydrochloric | $HCl$ |
| Hydrobromic | $HBr$ |
| Hydriodic | $HI$ |
| Tetraphenylboric | $H(C_6H_5)_4B$ |
| Nitric[b] | $HONO_2$ |
| Thiocyanic[b] | $HSCN$ |

[a]Sulfuric acid is strong only in the first step of dissociation; $HOSO_3^-$ has a $K_a$ of $1.1 \times 10^{-2}$.
[b]These acids appear strong in reasonably dilute solutions, but have been shown to have measurable, though large, $K_a$'s.

To evaluate the two terms on the right-hand side of 2-11, we may substitute solutions from 2-10* and 2-4:

$$[H_3O^+] = C + \frac{Q_w}{[H_3O^+]} \qquad (2\text{-}12^*)$$

Equation 2-12* is a quadratic equation in $H_3O^+$, and we may always solve it by means of the quadratic formula. However, before we do this, it would be well to investigate the possibility that one or the other term of the right-hand side of 2-12* may so predominate that we can simplify matters by discarding the other.

For example, suppose $C$ is a substantial number like $1 \times 10^{-2}$ F. Then from Eq. 2-12*, $[H_3O^+]$ is $1 \times 10^{-2}$ M, and $Q_w/[H_3O^+]$ is no larger than $10^{-12}$ M. It is certainly doing no violence to Eq. 2-12 to discard this relatively small term, and we see that in this case Eq. 2-10* is an excellent approximation.

On the other hand, as $C$ becomes very small, $Q_w/[H_3O^+]$ will approach $1 \times 10^{-7}$ M, and if $C$ becomes appreciably less than this limit, the second term in Eq. 2-12* predominates. The equation then approaches Eq. 2-6, which is certainly reasonable. One could hardly expect that the limit of the solution's composition as $C$ approaches zero could be other than that of pure water! In the range $10^{-6} > C > 10^{-8}$, neither simple solution applies accurately, and an exact solution must be found using Eq. 2-12*. It is surprising how rarely such cases arise in real life.

## pH in Solutions of Strong Bases

The question of the strength of a substance as a Brønsted base — that is, the strength with which it binds an additional proton — is fundamentally the same as that of the strengths of Brønsted acids, discussed above. If a molecule is a weak acid, its conjugate base is a relatively strong base. For the reaction of a base with water,

$$B^n + H_2O = BH^{n+1} + OH^-$$    (2-13)

The characteristic equilibrium quotient is given the symbol $Q_b$:

$$Q_b = \frac{[BH^{n+1}][OH^-]}{[B^n]}$$

and there is a simple relationship between the $Q_a$ (or $K_a$) of any acid and the $Q_b$ (or $K_b$) *of its conjugate base:*

$$Q_a Q_b = Q_w$$
$$K_a K_b = K_w$$    (2-14)

EXERCISE 2.1:   Prove the above relationships from the definitions.

It is sometimes asserted that if an acid is weak, its conjugate base must be strong. This is only relatively true; clearly there is an infinite number of pairs of numbers, both of which are small, whose product is the relatively tiny number $10^{-14}$. For example, the $K_a$ of the ion $H_2PO_4^-$ is roughly $10^{-7}$. From Eq. 2-14, the $K_b$ of its conjugate base $HPO_4^-$ is also $10^{-7}$.

For strong bases, then, we look to the conjugate bases of extremely weak acids. Examples are the alkoxide ions ($H_3CO^-$, $H_3CCH_2O^-$, etc.), amides ($NH_2^-$ and its derivatives), and the conjugate base of water, $OH^-$. Indeed, when any base stronger than $OH^-$ is dissolved in water, Eq. 2-13 proceeds far to the right, and an equivalent quantity of hydroxide ions is produced in the solution.

If we now ask what quantity of $OH^-$ is produced in solution by $C F$ strong base, we find that, algebraically, we are addressing the same problem that we have solved for strong acids. In view of the fact that equilibrium 2-13 lies far to the right, analogous to 2-8 for a strong acid, we may derive an approximate relation

$$[OH^-] = C$$    (2-15*)

and a more exact one

$$[OH^-] = C + \frac{Q_w}{OH^-}$$    (2-16*)

in a manner analogous to the derivations of (2-10*) and (2-12*).

EXERCISE 2.2:   Derive relations 2-15* and 2-16*.

The remarks following the derivation of Eq. 2-12* about the relative usefulness of Eqs. 2-10* and 2.12* apply similarly to Eqs. 2-15* and 2-16*.

## Weak Acids and Bases: Graphical Treatments

If the acid in question is not a categorically strong one, and is only partly dissociated in water, a new set of problems arises. Fortunately, there is a new tool for solving the problems: the value of $Q_a$, which is inaccessible in the case of a strong acid. With about equal frequency, one wishes to know the pH, knowing $C$ for a weak acid, or knows the pH (either through an instrumental measurement or because it is fixed by some other equilibrium) and wishes to calculate what fraction of a given acid-base pair is present in each of its conjugate forms. Since the latter calculation follows more directly from the definition of the $Q_a$, we will begin there.

**Distribution diagrams for conjugate acid-base systems.**   In a solution containing a weak acid and more or less of its conjugate base, there is a simple relation between the $[H^+]$ of the solution and the ratio of the concentrations of the weak acid and its conjugate base. Suppose we have a weak acid HX whose conjugate base we will call X. (When speaking about acids and bases in general, we will omit the charges since, unless some argument depends specifically on the electrical charge, it is better not to get in the habit of thinking of all acids as being electrically neutral, and all bases as negative. Whatever the charge on HX, that on X will clearly be one unit less positive.) The $Q_a$ (Eq. 2-9) may be rearranged to get at the ratio of the two forms:

$$\frac{[HX]}{[X]} = \frac{[H^+]}{Q_a} \qquad (2\text{-}17)$$

Since we may want to look at a wide range of acidities, we will convert to "p" notation by taking the negative logarithms of both sides:

$$-\log\left(\frac{[HX]}{[X]}\right) = pH - pQ_a \qquad (2\text{-}18)$$

Either Eq. 2-17 or Eq. 2-18 allows us to calculate the ratio of [HX] to [X] as a function of the known acidity of the solution.

Let us look at the same notion from another angle: What fraction of a given total quantity of HX and X is present as each form in a solution of a given pH? This kind of question arises fairly often in studies of acid-base equilibria in biological, physiological, or analytical chemistry. Suppose a constant total concentration $C$ of a substance is present in a solution whose pH we can vary independently. Regardless of the ratio of the two forms,

$$C = [HX] + [X] \qquad (2\text{-}19)$$

Suppose we call the ratio of the two forms $R$:

$$R = \frac{[HX]}{[X]}$$

Solving this equation for [X] and substituting the result in Eq. 2.19, we find

$$C = [HX] + \frac{[HX]}{R}$$

$$C = [HX]\left(1 + \frac{1}{R}\right)$$

$$\frac{[HX]}{C} = \frac{R}{(R+1)} \qquad (2\text{-}20)$$

The left-hand side of Eq. 2-20 is the fraction of the total quantity of the conjugate pair which is present as the acid form. Since the sum of the fractions present as HX and X must clearly be 1, we conclude fairly quickly that

$$\frac{[X]}{C} = 1 - \frac{[HX]}{C}$$

$$\frac{[X]}{C} = 1 - \frac{R}{R+1}$$

$$= \frac{1}{R+1} \qquad (2\text{-}21)$$

The fraction of any substance present as some particular form is sometimes given the symbol $\alpha$. Thus,

$$\alpha_{HX} = \frac{[HX]}{C}$$

and

$$\alpha_X = \frac{[X]}{C}$$

EXERCISE 2.3:   By combining Eqs. 2-17, 2-20, and 2-21, show that $\alpha_{HX} = [H^+]/([H^+] + Q_a)$ and $\alpha_X = Q_a/([H^+] + Q_a)$. These expressions will be useful in what follows.

A third way of expressing the ideas of Eqs. 2-19–2-21 is the fractional distribution diagram, which is a graph of $\alpha_{HX}$ and $\alpha_X$ *vs.* pH of the solution (Fig. 2-3). From Eq. 2-18, the important thing is not the pH of the solution alone, but the difference between the pH and the $pQ_a$ of our particular acid. Thus, the horizontal axis in Fig. 2-3 is graduated in units of pH relative to the $pQ_a$ of HX. In contemplating Fig. 2-3, you should picture the way the ratio $R$ and the two fractions $\alpha_{HX}$ and $\alpha_X$ change as the pH increases or decreases. What is the significance of the crossing point?

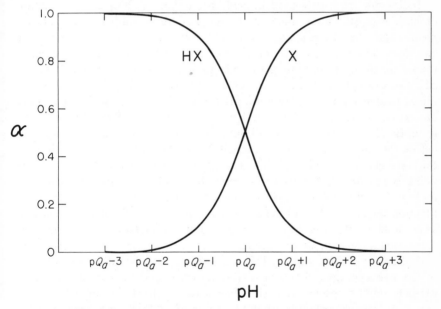

Fig. 2-3. Distribution diagram: $\alpha$ *vs.* pH for HX and its conjugate base X. pH is measured relative to $pQ_a$ of HX.

EXAMPLE 2.1:   A standard method for the analysis of organic material for total nitrogen content consists of "digesting" the sample with sulfuric acid to convert all nitrogen to $NH_4^+$, and then making the solution basic enough to convert the $NH_4^+$ to $NH_3$, which is distilled from the sample flask into a previously standardized solution of hydrochloric acid. If it is desired to convert 99.9% of the $NH_4^+$ to $NH_3$ before distillation, what must the pH of the solution be? The $pQ_a$ of $NH_4^+$ is about 9.3. A glance at Eq. 2-15 or at Fig. 2-3 indicates that we should expect that the pH will be several units higher than the $pQ_a$ of $NH_4^+$. We want the ratio $R$ to be 1/999; then from Eq. 2-16, $pH = pQ_a - \log(1/999)$, or $pH = 12.3$.

**Indicators.** An acid-base indicator is an acid-conjugate base pair the acid form of which has a different color from the basic form. When the indicator is present in solution, the color of the solution is a direct indication of $R$, the ratio of the concentrations of the acidic to the basic form of the indicator. Knowing $Q_a$, one may deduce the [H$^+$] in the solution more or less accurately depending on the accuracy with which $Q_a$ is known and the care with which the color is measured. At one extreme, with an approximate $Q_a$ and an eyeball measurement of the color, indicators provide a quick, useful estimate of the pH; at the other, with a carefully predetermined value of $Q_a$ and an instrumental (spectrophotometric) measurement of the ratio of the two colored forms, the calculated [H$^+$] *may* be very accurate indeed, though the ultimate limit on the accuracy is generally the uncertainty in the value of $Q_a$.

Indicators are widely used in acid-base titrations, because the reaction of equivalent quantities of acid and base (a condition called the *equivalence point*) results in a large change in the pH of the solution. If the indicator is so chosen that its p$Q_a$ nearly coincides with the pH at the equivalence point, then $R$ for the indicator changes by a large factor during the rapid change of pH. For example, in the titration of an acid with a base, $R$ is large before the equivalence point, because the pH is several units below the p$Q_a$ of the indicator; as the pH reaches and then exceeds the p$Q_a$, $R$ rapidly decreases to unity and then to a very small number. This is equivalent to saying that before the equivalence point is reached the predominant form of the indicator is the acid form; this rapidly changes to the basic form as the equivalence point is reached and passed. The operator's subjective judgement of the midpoint of this color change is called the *end point*, and for an accurate titration the end point and the equivalence point should coincide as nearly as possible. In titrations as in any application of indicators for pH measurement, it is important that the indicator not change the pH of the solution in the process of measuring it. This will be the case if the total concentration of the two conjugate forms is very small compared to that of any other pH-determining species. It is therefore a requirement, fortunately met by many indicator substances, that the color of at least one form be very intense.

So-called "universal indicators" are mixtures of indicators with p$Q_a$'s so spaced that the color of the solution changes continuously over a wide range of pH. Since at any particular pH a complex mixture of colors is present, a direct calculation of the pH from an instrumental color measurement is difficult, and color-matching charts are provided with the indicator. With good charts, well-chosen indicators, and ideal conditions of solution buffering, lighting, and operator color sensitivity, it is possible to estimate the pH of a solution to within ± about 0.2 units. This corresponds to an uncertainty in [H$_3$O$^+$] of about 60% of its value. "pH papers" are strips of presumably unreactive filter paper impregnated with universal indicator.

Narrow-range papers, which cover a range of 1 to 2 units, allow estimation of the pH to within 0.1 pH unit, provided that the $pQ_a$'s assumed in the color-matching chart for the paper are valid for the solution being tested. This uncertainty in pH corresponds to $\pm 25\%$ in [H$^+$]. Since there exist more accurate means of measuring pH over a wide range, the only advantage pH papers or universal indicators offer is convenience, and they are not generally used for serious measurements unless the allowable margin for error is very broad.

**Logarithmic distribution graphs.**    Although graphs such as that in Fig. 2-3 give a good qualitative picture of the distribution of an acid-base pair among its conjugate forms, they are not particularly helpful in solving quantitative problems, partly because they are unreadable at very large or very small values of $R$, and partly because in many situations, one is interested in the absolute concentrations of the two forms rather than their relative concentrations $\alpha$. The problem of very large and very small $R$ is best solved by converting the graph to a logarithmic one. Logarithmic distribution graphs have, as we will see, the additional advantage that they can be drawn accurately without calculating each point.

The logical beginning to a derivation of a logarithmic distribution graph is to put Eqs. 2-20 and 2-21 in logarithmic form:

$$\log \alpha_{\text{HX}} = \log R - \log (R + 1) \qquad (2\text{-}22)$$

$$\log \alpha_{\text{X}} = \log 1 - \log (R + 1)$$
$$= -\log (R + 1) \qquad (2\text{-}23)$$

The log $(R + 1)$ terms cannot be expanded to any simpler form, but we are deriving this kind of graph to accommodate very large and very small $R$, so that we may logically investigate the limiting forms of 2-22 and 2-23 under those conditions. From Eq. 2-18, this will be true at pH's relatively distant from the $pQ_a$ of HX.

**Limiting conditions at large $R$.**    If $R$ is much greater than 1, Eq. 2-22 approaches log $\alpha_{\text{HX}} = 0$. This very simple equation has a very simple graph (see Fig. 2-4a). Equation 2-23 approaches log $\alpha_{\text{X}} = -\log R$, and in view of Eq. 2-18, this is

$$\log \alpha_{\text{X}} = \text{pH} - pQ_a \qquad (2\text{-}24^*)$$

A graph of Eq. 2-24* on log $\alpha$, pH coordinates is a straight line with slope $+1$ and intercept $-pQ_a$ (Fig. 2-4a).

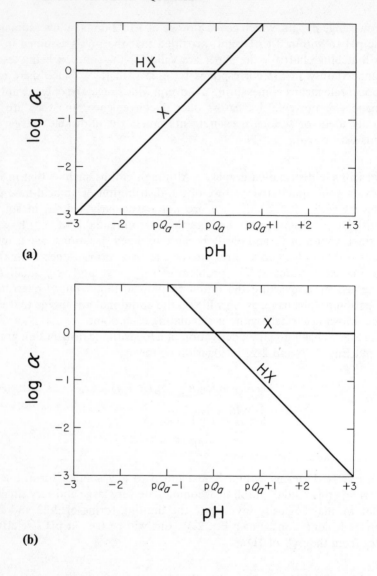

**Fig. 2-4. (a)** Limiting lines for log $\alpha$ at $R \gg 1$. **(b)** Limiting lines for log $\alpha$ at $R \ll 1$.

**Limiting conditions at small $R$.**    If $R$ is much less than 1, Eq. 2-22 approaches the limit log $\alpha_{HX} = \log R$, or

$$\log \alpha_{HX} = pQ_a - pH \qquad (2\text{-}25^*)$$

and Eq. 2-23 approaches log $\alpha_X = 0$, again a simple horizontal straight line. The graphs of these two limits are shown in Fig. 2-4b.

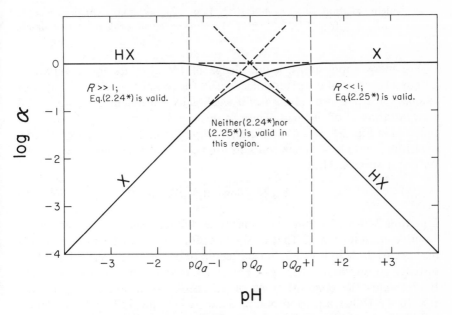

**Fig. 2-5.** Complete graph of log $\alpha$ *vs.* pH for the HX-X pair.
The intersection of the dotted lines is the system point.

Since the two limiting conditions discussed above are mutually contra-
dictory, the equations resulting from them can only be graphed together by
indicating limits to the domains within which the two approximations
apply. This is done in Fig. 2-5.

In the awkward region within which neither limiting condition applies, we
are forced to use Eqs. 2-22 and 2-23 to plot log $\alpha$ for HX and X. However,
since this region is relatively short, and since we know that the nonlinear
graph must connect the two straight-line portions of a given log $\alpha$ line, a
single calculated point will suffice. From Eq. 2-18 or Fig. 2-3, both $\alpha$'s
are 0.5 at pH = $pQ_a$. Then both log $\alpha$'s will be log 0.5, or $-0.30$. That is,
log $\alpha_{HX}$ will cross log $\alpha_X$ at the point pH = $pQ_a$ and log $\alpha = -0.30$. With
this point as a guide, it is not difficult to sketch in the curved portion of
both log $\alpha$ lines, as indicated in Fig. 2-5.

Let us consider Fig. 2-5 before modifying it to reflect absolute rather
than relative concentrations. Note that very small values of $\alpha$ for both
species are easily read from this graph, because their logarithms vary
linearly with pH. Values of $\alpha$ close to unity are no easier to read than on a
linear distribution graph like Fig. 2-3; however, since $\alpha_{HX} + \alpha_X = 1$,
these are easily deduced from the value of $\alpha$ for the conjugate form.

EXAMPLE 2.2:   Solve the problem of example 2.1 using Fig. 2-5. The desired
condition is that $\alpha_{NH_3} = 0.999$. We find the necessary pH by looking for the

equivalent solution $\alpha_{NH_4^+} = 1 \times 10^{-3}$. Log $\alpha_{NH_4^+} = -3.0$, which is true at pH = $pK_a + 3$, from Fig. 2-5. This is the same as the algebraic result of example 2.1.

For many applications, a knowledge of the $\alpha$ values for a particular pair as a function of the pH is not sufficient for complete understanding. For example, the pH of a solution of a weak acid is a function of the formal concentration $C$ of the acid. There is no way for a knowledge of $C$ to be reflected in Fig. 2-5. It is not difficult, however, to convert a log $\alpha$ graph to a logarithmic graph of molar concentrations. From its definition, $\alpha = M/C$ for either species. Then

$$\log M = \log \alpha + \log C \qquad (2\text{-}26)$$

Equation 2-26 states that log molarity is different from log $\alpha$ only by an additive constant log $C$. That is, the lines for each species will be identical in shape to those in Fig. 2-5, but shifted up or down on the vertical axis by a distance corresponding to log $C$. Figure 2-6 is an example of such a graph. It represents the concentrations of all species in a solution containing $1 \times 10^{-2}$ F HOCl, a monoprotic weak acid whose $pK_a$ is 7.30. Note that on a log concentration graph we can include concentrations of species other than the members of the particular conjugate pair we are studying. The lines for $H_3O^+$ and $OH^-$ are derived from the relationships log [$H_3O^+$] = $-$pH and log [$OH^-$] = $-$pOH = pH $- pQ_w$. These lines are most easily located by noting that they have slopes of $-1$ and $+1$, respectively, and that they cross at pH 7.0.

EXERCISE 2.4:   Prove the statements in the preceding sentence.

Let us now use Fig. 2-6 to deduce the pH of the solution which it represents: $1 \times 10^{-2}$ F HOCl. A proton balance relative to the reference level $H_2O$ + HOCl yields

$$[H_3O^+] = [OCl^-] + [OH^-] \qquad (2\text{-}27)$$

A logarithmic concentration diagram affords a method of determining the important terms in equations like 2-27. Note that in Fig. 2-6, the left-hand side of Eq. 2-27 is represented by a line of slope $-1$, and the right-hand side by lines of slope either $+1$ or zero. There must be intersections, then, at which [$H^+$] equals one of the terms on the right-hand side of Eq. 2-27. If we look for the highest such intersection, we will thereby find the larger of the two terms in the right-hand side of Eq. 2-27. In Fig. 2-6, this is the intersection of the $H^+$ line with the $OCl^-$ line at pH = 4.65. If we note that at that pH, the $OH^-$ line is at log [$OH^-$] = $-9.35$, over four log units below the $OCl^-$ line, we see that the contribution of [$OH^-$] to Eq. 2-27

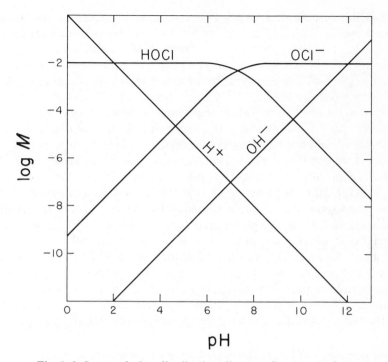

**Fig. 2-6.** Log molarity distribution diagram for $1 \times 10^{-2}$ $F$ hypochlorous acid.

is negligibly small. If log [OH$^-$] is less than log [OCl$^-$] by more than four units, then [OH$^-$] is less than [OCl$^-$] by a factor greater than $10^4$. The H$^+$-OCl$^-$ intersection then fulfills the proton balance condition to an excellent approximation, and the pH of the solution is 4.65.

To summarize rules for constructing a logarithmic distribution graph:

1. The H$^+$ and OH$^-$ lines have an invariant slope of $-1$ and $+1$, respectively, and intersect at pH 7.
2. At pH less than p$Q_a$ by at least 1 unit, the HX line is horizontal, and the X line has slope $+1$. Both lines, if extended to pH $=$ p$Q_a$, would intersect at a point (pH $=$ p$Q_a$, log $M$ $=$ log $C$) called the *system point*.
3. At pH greater than p$Q_a$ by at least 1 unit, the X line is horizontal, and the HX line has slope $-1$. Again, both lines, if extended back to pH $=$ p$Q_a$, would pass through the system point.
4. In the region between p$Q_a - 1$ and p$Q_a + 1$, the HX and X lines are curved, and intersect at a point 0.3 log units directly below the system point.

These rules may, as we shall see, be extended to the construction of logarithmic graphs for systems more complex than monoprotic acids.

You should now be asking what would have happened in our problem if the $H^+$-$OCl^-$ intersection had not been so far above the $OH^-$ line under the conditions of the problem. For example, if we had chosen a smaller value of C, the $H^+$-$OCl^-$ intersection would have approached the $OH^-$ line and it might not have been possible to discard the $[OH^-]$ term in Eq. 2-27. The answer is that by taking the highest intersection of lines for members on opposite sides of a proton balance equation, we will certainly never make a gross mistake. The worst that could happen would be that two lines on the same side of a proton balance would coincide. To ignore one in favor of the other, as we did in the above problem, would be to make an error of a factor of two. In that problem, the pH would have been different from the simple graphical solution by $\log \sqrt{2}$, or 0.15 pH units. Two comments on this result are appropriate. First, it is frequently the case that 0.15 pH units represents an entirely acceptable error, either because only a rough answer is sought, or because the $pQ_a$ of the acid is not accurately known in the first place. Second, if an accurate calculation of the pH is desired, an algebraic calculation can be done much more quickly and conveniently if order-of-magnitude values of the concentrations of all species are known before beginning. This a logarithmic graph can give even when, because two intersections coincide, an accurate graphical solution is not possible.

**Weak bases.** The graphical treatment of an acid-base pair as a single chemical system makes unnecessary the nonfundamental distinction between problems involving weak acids and those involving weak bases. Given the proper proton balance equation, a problem involving a solution of a weak base can be solved using the same distribution diagram as one involving a solution of its conjugate acid. It is true that some substances are known primarily as bases and others as acids, but the distinction rests only on the question of which member of the pair is electrically neutral. Thus the term "phenol" describes the electrically neutral acid member of the pair $C_6H_5OH$-$C_6H_5O^-$, and the term "pyridine" the electrically neutral base member of the pair $C_6H_5N$-$C_6H_5NH^+$, even though in these particular cases, the named member happens to be the less reactive in Brønsted equilibria. Conjugate pairs are identified by their electrically neutral member because that substance can be isolated as a pure compound (for example, pyridine is a fishy-smelling oil), whereas no single ion, as we have seen, can ever be isolated in visible quantities. When we speak of substances as solutes, however, there is no reason to distinguish between neutral acids such as phenol and charged ones such as $C_6H_5NH^+$.

Let us return to Fig. 2-6, and assume that we are interested in determining the pH of $1 \times 10^{-2}$ $F$ NaOCl. This substance is completely dissociated into its ions in water solutions, and the solution will also be $1 \times 10^{-2}$ $M$ in $OCl^-$. A proton balance using $H_2O$ and $OCl^-$ as reference levels produces

$$[HOCl] + [H^+] = [OH^-] \qquad (2\text{-}28)$$

We proceed as before to look for the highest intersection of lines representing species on opposite sides of this proton balance. There is an $OH^-$-HOCl intersection at pH 9.7, and at that pH the $H^+$ line is more than 5 units below the intersection. Then $[H^+] = [HOCl] \times 10^{-5}$, and we can justifiably ignore it in Eq. 2-28. The proton condition reduces to $[HOCl] = [OH^-]$, and this condition is represented by the HOCl-$OH^-$ intersection at pH 9.7. The pH of $1 \times 10^{-2}$ $F$ NaOCl is 9.7.

**Polyprotic systems: graphical solutions.** A large class of substances have more than one removable proton. Such substances are called polyprotic Brønsted acids or bases (the distinction between acidic and basic substances having the same validity as in monoprotic compounds). Aqueous solutions of these substances present a few new features which our study of monoprotic systems has prepared us to deal with.

First, let us define some equilibrium constants. Suppose we are interested in the polyprotic acid $H_nY$. It may donate protons, one at a time, in $n$ steps:

$$H_nY + H_2O = H_{n-1}Y + H_3O^+$$
$$H_{n-1}Y + H_2O = H_{n-2}Y + H_3O^+$$

etc.

We indicate the strengths of the acids $H_nY$, $H_{n-1}Y$, etc., by *stepwise* equilibrium quotients analogous to the $Q_a$'s of monoprotic acids:

$$Q_1 = \frac{[H^+][H_{n-1}Y]}{[H_nY]}$$

$$Q_2 = \frac{[H^+][H_{n-2}Y]}{[H_{n-1}Y]}$$

etc.

A list of some common and relatively important polyprotic acids is given in Table 2.3.

A quantitative study of polyprotic systems may begin with a distribution graph. Figures 2-7 and 2-8 are linear and logarithmic fractional distribution graphs, analogous to Figs. 2-3 and 2-5 for the triprotic acid $H_3PO_4$. Note that an additional crossing point appears for each additional constant. There are as many system points for a polyprotic acid as there are $K$'s. Between the system points are pH ranges over which a particular one of the four species predominates. At pH less than 2, the predominant form is $H_3PO_4$; between 3 and 7 it is $H_2PO_4^-$; between 8 and 11, $HPO_4^{--}$; and above 13 it is $PO_4^{3-}$. The only really new feature the logarithmic graph presents is that the slope of a given line changes each time that line passes

TABLE 2.3.  SOME POLYPROTIC ACIDS

| I. Inorganic Acids | $pK_1$ | $pK_2$ | $pK_3$ | $pK_4$ |
|---|---|---|---|---|
| $H_2CO_3 + CO_2$ | 6.35 | 10.33 | | |
| $H_2CO_3$ | 3.88 | 10.33 | | |
| $H_3PO_4$ | 2.172 | 7.211 | 12.360 | |
| $H_4P_2O_7$ | 1.52 | 2.36 | 6.60 | 9.25 |
| $H_3AsO_4$ | 2.19 | 6.94 | 11.50 | |
| $H_2S$ | 6.99 | 12.89 | | |
| $H_2SO_3$ | 1.76 | 7.20 | | |
| $H_2Se$ | 3.89 | 11.0 | | |

| II. Carboxylic Acids | $pK_1$ | $pK_2$ | $pK_3$ |
|---|---|---|---|
| Oxalic | 1.25 | 4.28 | |
| Malonic | 2.85 | 5.66 | |
| Maleic | 1.92 | 6.22 | |
| Fumaric | 3.02 | 4.39 | |
| Succinic | 4.19 | 5.48 | |
| Tartaric | 3.46 | 5.05 | |
| Citric | 3.128 | 4.761 | 6.395 |
| o-Phthalic | 3.10 | 5.40 | |
| m-Phthalic | 2.30 | 4.66 | |

| III. Amino Acids | $pK_1$ | $pK_2$ | $pK_3$ |
|---|---|---|---|
| Glycine (+1)[a] | 2.35 | 9.78 | |
| Alanine (+1) | 2.34 | 9.87 | |
| Leucine (+1) | 2.36 | 9.6 | |
| Tyrosine (+1) | 2.20 | 9.19 | 10.43 |
| Aspartic acid (+1) | 2.1 | 3.9 | 10.0 |
| Histidine (+2) | 1.8 | 6.1 | 9.2 |

[a]Charge on the parent acid.

a system point. For example, the slope of the $H_3PO_4$ line changes from $-1$ to $-2$ as the pH increases from about 6 to about 9, and from $-2$ to $-3$ as the pH increases from 11 to 13. The reason for this behavior is not difficult to see. $\alpha_{H_3PO_4}$ is $[H_3PO_4]/C$, and in view of the predominance of each species in a given range of pH, $\alpha_{H_3PO_4}$ has the following simplified forms (the symbol $\alpha_3 = \alpha_{H_3PO_4}$ is used for simplicity):

$$pH < pQ_1: \qquad C = [H_3PO_4] \quad \text{and} \quad \alpha_3 = 1 \qquad (2\text{-}29^*)$$

$$pQ_1 < pH < pQ_2: \qquad C = [H_2PO_4^-]$$

$$\alpha_3 = \frac{[H_3PO_4]}{[H_2PO_4^-]}$$

$$= \frac{[H^+]}{Q_1}$$

$$\log \alpha_3 = pQ_1 - pH \qquad (2\text{-}30^*)$$

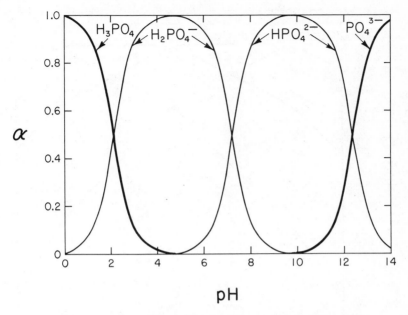

**Fig. 2-7.** $\alpha$ *vs.* pH for $H_3PO_4$ and its conjugate bases. The lines for $H_3PO_4$ and $HPO_4^{--}$ appear to join, as do those for $H_2PO_4^-$ and $PO_4^{-3}$. This is an artifact; they cross at very small $\alpha$ values. See Fig. 2-8.

and the slope of log $\alpha_3$ is $-1$, just as in the case of a monoprotic acid, which we have considered. But if

$$pQ_2 < pH < pQ_3: \qquad C = [HPO_4^{2-}]$$

$$\alpha_3 = \frac{[H_3PO_4]}{[HPO_4^{2-}]}$$

$$= \frac{[H^+]^2}{Q_1 Q_2}$$

$$\log \alpha_3 = pQ_1 + pQ_2 - 2pH \qquad (2\text{-}31^*)$$

Finally, when

$$pH > pQ_3: \qquad C = [PO_4^{3-}]$$

$$\alpha_3 = \frac{[H_3PO_4]}{[PO_4^{3-}]}$$

$$= \frac{[H^+]}{Q_1 Q_2 Q_3}$$

$$\log \alpha_3 = pQ_1 + pQ_2 + pQ_3 - 3pH \qquad (2\text{-}32^*)$$

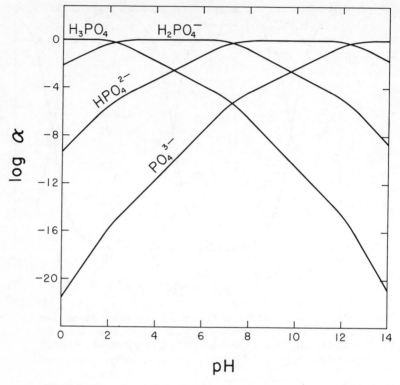

**Fig. 2-8.** Log fractional distribution diagram for $H_3PO_4$.

These approximate equations for log $\alpha_3$ are represented on a log distribution graph by lines of slope 0, −1, −2, and −3, respectively.

We again easily convert the fractional logarithmic graph to a log concentration graph by displacing each line a vertical distance equal to log $C$ for any chosen value of $C$. Figure 2-9 gives the result for 0.1 $F$ $H_3PO_4$.

If we use Fig. 2-9 as we have the distribution diagram for the monoprotic system, we will find some familiar features reappearing in this polyprotic system, and a few new ones. For example, suppose we consider the question of the pH of a 0.1 $F$ solution of $H_3PO_4$. This situation will differ from the monoprotic case only to the extent that the second and third protons on $H_3PO_4$ are donated to water. A proton balance from the reference level $H_3PO_4 + H_2O$ is

$$[H^+] = [OH^-] + [H_2PO_4^-] + 2[HPO_4^{--}] + 3[PO_4^{3-}]$$

The highest intersection of $[H^+]$ with any term on the right-hand side of the proton balance is the $H^+$-$H_2PO_4^-$ intersection at pH 1.6. At that point, all of the other terms on the right-hand side of the proton balance are negligible;

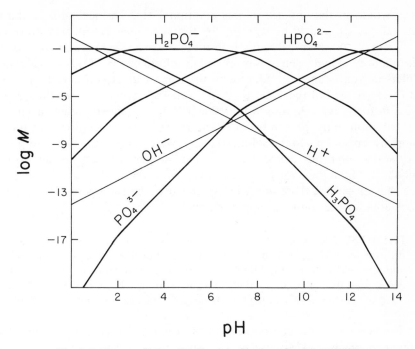

**Fig. 2-9.** Log molarity distribution diagram for 0.1 $F$ $H_3PO_4$.

the largest, $[HPO_4^-]$, is smaller than $[H_2PO_4^-]$ by a factor of approximately $10^5$, and the others are smaller still. Consequently, the $H^+$-$H_2PO_4^-$ intersection satisfies the proton balance to an excellent approximation, and no other terms need be considered. The pH of the solution is 1.6. Now let us consider the significance of this result. The fact that $HPO_4^{2-}$, $PO_4^{3-}$, and $OH^-$ are negligible in the proton balance means that essentially all the hydrogen ions in solution came from the first ionization of $H_3PO_4$. The second and third steps might as well not exist. Before we gloat too much over the mathematical simplification which this result introduces, we should ask whether it makes sense. Why is a triprotic acid behaving no differently from a monoprotic one? The answer comes from a consideration of the proton-transfer equilibria:

$$H_2PO_4^- + H_2O = HPO_4^{--} + H_3O^+$$
$$HPO_4^{--} + H_2O = PO_4^{3-} + H_3O^+$$

These represent the dissociations of what are to begin with quite weak acids: their $Q_a$'s ($Q_2$ and $Q_3$ of $H_3PO_4$, respectively) are roughly $10^{-7}$ and $10^{-12}$. Furthermore, there is in solution a good deal of $H^+$ supplied by the first dissociation. The above two equilibria lie farther to the left than they would in the absence of this additional $H^+$. The second and third ionizations are said to be *repressed* by the additional $H^+$ already present.

We may ask whether the simple result of the preceding paragraph — that the quantity of hydrogen ions produced in solution by $H_3PO_4$ is virtually no different from what is produced by the same formal concentration of a monoprotic acid of the same $Q_a$ — is a generally applicable conclusion. If we consider Fig. 2-9, we see that the $[HPO_4^-]$ would represent a larger contribution to the proton balance (and the second dissociation contribute more hydrogen ions to the solution) if $pQ_2$ were numerically closer to $pQ_1$. For some polyprotic acids, it is true that the gap between the successive $pK$'s is less than the 5 log units of $H_3PO_4$. However, from Table 2.3 we can see that in general there is a wide spread between successive dissociation constants for the common polyprotic acids.

EXERCISE 2.5:    From the data in Table 2.3, construct a log concentration diagram for 0.1 $F$ oxalic acid. Does the $H^+$ contributed by the second dissociation add appreciably to the total $[H^+]$ in the solution? Approximately how close together must the first and second pK of a polyprotic acid be for the second dissociation to contribute 10% of the total hydrogen ion concentration of the solution?

The generally large gap between the successive dissociation constants of polyprotic acids makes sense on the basis of the structural influences on $K_a$'s discussed earlier in this chapter. When an acid dissociates, its electrical charge decreases by one unit. The changed electrical environment of the second ionizable proton resulting from this decrease in charge makes its removal a very different matter from the removal of the first. Naturally, for acids in which the site of the removal of the first proton is distant from the location of the second proton, the effect of this decrease in charge is relatively less. For oxalic, malonic, and succinic acids, there are zero, one, and two $CH_2$ groups, respectively, separating the two carboxylate groups. Note in Table 2.3 the effect this has on the difference between $pK_1$ and $pK_2$ in this series of acids. Interestingly, it can be shown on statistical grounds that the minimum ratio of $K_1$ to $K_2$ for a diprotic acid is 4, *provided* that the two dissociation sites are identical and are symmetrically placed on the molecule. Obviously, this limit (which is equivalent to a difference in $pK$'s of log 4, or 0.6) is only approached for acids with widely separated acidic sites. For example, in azelaic acid, $HOOC(CH_2)_7COOH$, $pK_1 = 4.55$, $pK_2 = 5.41$, and this separation of $pK$'s amounts to a ratio $K_1/K_2 = 7.2$.

**Amphiprotic substances.**    The ions $H_2PO_4^-$ and $HPO_4^{--}$ display a property we have not yet encountered: they are simultaneously Brønsted bases and acids. For example, $H_2PO_4^-$ can lose a proton to form $HPO_4^{--}$ or gain one to form $H_3PO_4$. Substances that have this property are called *amphiprotic.* Obviously, the conjugate base of any polyprotic acid (or the conjugate acid of any polyprotic base) is amphiprotic.

Let us consider what the effect on the pH of pure water will be if an amphiprotic substance is dissolved in it. For example, what is the pH of 0.1 $F$ $NaH_2PO_4$, assuming complete dissociation of this salt into $Na^+$ and $H_2PO_4^-$ ions? The answer to this question may be read from the logarithmic distribution graph, Fig. 2-9, if we adjust the proton balance to fit the conditions of the problem. If we imagine that the solution was prepared by adding $NaH_2PO_4$ to $H_2O$ these substances should form the reference level for a proton balance which then looks like

$$[H^+] + [H_3PO_4] = [OH^-] + [HPO_4^{--}] + 2[PO_4^{3-}] \qquad (2\text{-}33)$$

The coefficient 2 in front of $[PO_4^{3-}]$ should be comprehensible by this time. Every mole of $PO_4^{3-}$ in solution represents the absence of two moles of protons from a species of the reference level. If we now seek in Fig. 2-9 the highest intersection of any term on the left side of the proton balance with any term on the right, we find the $H_3PO_4$-$HPO_4^{2-}$ intersection at pH 4.7. At this pH no other term in the proton-balance equation is appreciable compared to $[H_3PO_4]$ and $[HPO_4^{--}]$, except $[H^+]$, which is smaller than $[H_3PO_4]$ by about 1.5 log units, representing a factor of about 30. That is, 97% of the left-hand (excess) side of the proton balance is represented by $[H_3PO_4]$ and only about 3% by $[H^+]$. We may justifiably ignore the $[H^+]$ term for a good approximate solution: the pH of the solution is 4.7.

The result of this consideration, that the simplified proton balance for a solution of $H_2PO_4^-$ in water is

$$[H_3PO_4] = [HPO_4^{--}] \qquad (2\text{-}34^*)$$

may be generalized as a first approximation for any amphiprotic substance: roughly speaking, as many amphiprotic molecules gain protons as lose them, resulting in an approximate proton balance in which the concentration of the conjugate acid of the amphiprotic substance equals the concentration of its conjugate base. This is true (approximately) despite the discrepancy between the inherent acidity and basicity of $H_2PO_4^-$. As an acid, $H_2PO_4^-$ has a $K_a$ ($= K_2$ for $H_3PO_4$) of $6 \times 10^{-8}$, and as a base, it has a $K_b$ ($= K_w/K_1$) of $1.3 \times 10^{-12}$, smaller than its $K_a$ by a factor of over $10^4$. We will return to this result when we consider an algebraic treatment of the problem. For now, we may conclude by inspection of Fig. 2-9 that our approximate proton balance (Eq. 2-34*) will be a good approximation as long as the conjugate acid-conjugate base intersection is safely above any line representing another species in the proton balance equation. Usually this means that the approximation will be valid as long as the formal concentration of the amphiprotic substance is large enough to keep the conjugate acid and conjugate base lines above the $H^+$ and $OH^-$ lines.

EXERCISE 2.6: Draw a logarithmic distribution diagram for 0.01 $F$ phosphoric acid. How valid is the approximate proton balance discussed above for a solution of 0.01 $F$ $H_2PO_4^-$? For 0.01 $F$ $HPO_4^{2-}$? How might the behavior of $HPO_4^-$ in this case be better described?

## Weak Acids and Bases: Algebraic Treatments

Graphical treatments of acid-base equilibria are convenient, and often the only possible solution to complex problems. However, they have two disadvantages. First, they are necessarily approximate, because the accuracy with which a number may be determined from a graph of its logarithm is far less than that with which it may be calculated directly. Second, since a log graph such as Fig. 2-6 is based on concentrations, it must be adjusted or redrawn for each new formal concentration considered. Also, paper, pencil, and a slide rule are more portable and generally available than graph paper, and for simple problems it is much easier to perform a direct calculation of the desired quantity than to construct a graph.

These considerations make it worthwhile to develop some algebraic techniques for analyzing equilibrium systems. An algebraic solution to a chemical problem is no different from the algebra which we all encountered in junior high school. Only if there exist as many independent equations as there are unknowns in the system is a complete algebraic solution possible. We will consider first techniques for meeting this purely mathematical requirement, and then examine various approximations which will simplify the solution of the resulting equations. To put the approach less formally, we will see first how bad things can be and then consider ways for cheating a little. As long as we bear in mind carefully the limits that approximate algebraic solutions put on the accuracy of our calculations, there is no reason not to take this approach. In fact, it is usually the case that the numbers — equilibrium quotients and formal concentrations — that are fed into an algebraic solution have generous margins of uncertainty which do not justify a very precise solution.

**Equilibrium quotients.** The obvious place to begin looking for the independent equations we need to describe our system is with equilibrium quotients. For every equilibrium in the solution, the equilibrium quotient requires that a definite relationship exist between the concentrations of the products and reactants in that equilibrium. Some equilibrium quotients may be found to be derivable from others, and thus, of course, do not constitute *independent* equations for the purpose of algebraic solutions. For example, the diprotic acid $H_2S$ has two independent equilibrium quotients, $Q_1$ and $Q_2$. We might also include the overall dissociation quotient, defined as

$$Q = \frac{[H^+]^2[S^{--}]}{[H_2S]} \qquad (2\text{-}35)$$

This is the equilibrium quotient for the reaction

$$2H_2O + H_2S = S^{--} + 2H_3O^+ \qquad (2\text{-}36)$$

However, just as 2-36 is the sum of the two stepwise dissociations of $H_2S$, Eq. 2-35 is the product of the stepwise dissociation quotients, $Q_1Q_2$.

**Other eternal relationships.**    In dealing with the autoprotolysis equilibrium of water, we found that stoichiometry, charge balance, and proton balance can yield information on equilibrium systems. Though they were redundant in the case of autoprotolysis, these relationships may be independent in more complex problems.

**Mass balance.**    Once we put a given quantity of a substance into a solution, the atoms composing that substance do not change in number, and we may be sure that, despite any reaction that may occur, the total quantity of a given atom (or unreacting group, such as $NO_3^-$ or $SO_4^{--}$) remains constant. For example, consider a weak diprotic acid such as oxalic acid, HOOCCOOH. If we prepare a solution $C\ F$ in oxalic acid, then

$$C = [HOOCCOOH] + [HOOCCOO^-] + [^-OOCCOO^-] \qquad (2\text{-}37)$$

That is, regardless of how the substance distributes itself among its three forms, their total concentration must remain constant if the $C_2O_4^{--}$ group is to be neither created nor destroyed.

Let us now turn to the problems we have already solved graphically, and see how we can produce algebraic solutions.

**Weak acids: algebraic treatment.**    Consider a solution $C\ F$ in the weak acid HX. What is its pH? To answer this question without recourse to distribution graphs, we note that the solution contains four unknown concentrations, $[H_3O^+]$, $[OH^-]$, $[HX]$, and $[X]$, and we set up four independent equations:

$$Q_a = \frac{[H^+][X]}{[HX]} \qquad (2\text{-}38)$$

$$Q_w = [H^+][OH^-] \qquad (2\text{-}39)$$

mass balance:

$$C = [HX] + [X] \qquad (2\text{-}40)$$

and proton balance relative to the reference level $H_2O + HX$:

$$[H^+] = [X] + [OH^-] \qquad (2\text{-}41)$$

Let us assemble these four equations in such a way as to reduce them to one equation with one unknown, which must have at least one accessible solution. Since Eq. 2-38 relates four fundamental quantities, we will rearrange the other three equations for substitution into 2-38 as follows:

$$[OH^-] = \frac{Q_w}{[H^+]}$$

$$[HX] = C - [X]$$

$$[X] = [H^+] - [OH^-]$$

Equation 2-38 now becomes

$$Q_a = \frac{[H^+]([H^+] - [OH^-])}{C - [H^+] + [OH^-]} \qquad (2\text{-}42a)$$

$$Q_a = \frac{[H^+]([H^+] - Q_w/[H^+])}{C - [H^+] + Q_w/[H^+]} \qquad (2\text{-}42b)$$

Equation 2-42b is a single equation with one unknown and may in principle be solved for $[H^+]$. However, if you expand Eq. 2-42b, you will find that it is a cubic in $[H^+]$. Not only are there three solutions (only one of which will make sense as an answer to our problem), but the roots of a cubic equation are not generally found by a straightforward analytical procedure. You may have encountered successive-approximation methods for finding roots of equations of high degree, but you will recall that these methods, whether graphical or analytical, are tedious.

Before we try to find the exact solution of Eq. 2-42b, we should examine it carefully to see whether it might not be possible to make simplifying approximations. The best method for examining an entire complex equation at once is to graph it. Equation 2-42b contains three variables: $C$, $[H^+]$, and $Q_a$, the last depending on the identity of the acid under consideration. A complete graph would be three-dimensional; however, a two-dimensional graph with projected contour lines representing various values of the third variable is also satisfactory. Figure 2-10 shows pH as a function of log $C$

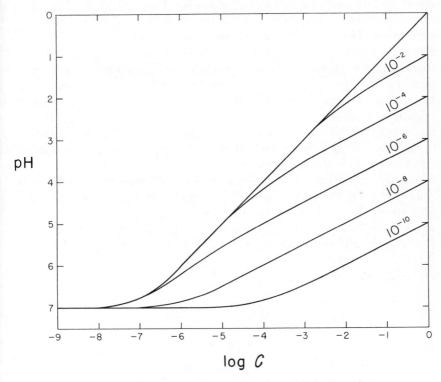

**Fig. 2-10.** pH *vs.* log $C$ for weak acids with indicated $Q_a$ calculated from Eq. 2-42b.

for various values of $Q_a$; points were obtained by rearranging 2-42b to the form:

$$C = \frac{[H^+]^3 + [H^+]^2 Q_a - Q_w[H^+] - Q_w Q_a}{Q_a[H^+]} \qquad (2\text{-}43)$$

and calculating $C$ consistent with various assumed values of $[H^+]$. This technique, a generally useful one, avoids the onerous job of solving a cubic equation for each plotted point. The results were converted to pH and log $C$, because logarithmic variables have the advantage of compressing a wide range of data into a small space, and because the graph assumes a pleasingly simple shape this way. Note first that as the solutions become more dilute (log $C$ becomes more negative) all lines approach the pH value 7; that is, regardless of the $Q_a$ value of an acid, a sufficiently dilute solution of it resembles pure water. Equations 2-42 and 2-43 have another simple limit: note that all lines representing acids with $Q_a$ larger than $10^{-6}$

merge with a diagonal line, whose equation is pH $= -\log C$. That equation is the logarithmic form of Eq. 2-10*, which gives the [H⁺] in a solution of a strong acid. The more dilute a solution of a weak acid, the more nearly it resembles a strong acid. This result also makes sense if we consider that, as a solution of a weak acid is made more dilute, collisions between hydronium ions and conjugate-base species become less likely, whereas chances that any particular acid molecule will encounter a water molecule remain essentially constant. The rightward direction of equilibrium 2-8 then is unaffected, but the leftward reaction is made less likely. Without a change in the value of $Q_a$, a larger fraction of the HX present in solution is dissociated as the formal concentration decreases.

Finally, between the two limiting lines discussed above is a region within which the pH of the solution does depend on the $Q_a$ of the weak acid, and, for a given value of $Q_a$, depends on log C in a simple-appearing way. We can simplify Eq. 2-42a considerably if we note that in this region, the pH is less than 7 by several units (that is, from Eq. 2-4, [H⁺] $\gg$ [OH⁻]), and that the pH is greater than $-\log C$ (that is, [H⁺] $\ll C$). Applying these conditions to Eq. 2-42a, we find

$$Q_a = \frac{[H^+]^2}{C} \qquad (2\text{-}44^*)$$

$$[H^+] = (Q_a C)^{1/2} \qquad (2\text{-}45^*)$$

$$pH = \tfrac{1}{2}(pQ_a - \log C) \qquad (2\text{-}46^*)$$

A graph of Eq. 2-46* on pH, $(-\log C)$ axes is a straight line with slope $+\tfrac{1}{2}$. Since the intercept of Eq. 2-46* is $pQ_a/2$, we obtain a set of parallel lines.

Our scrutiny of Fig. 2-10 leads us to the conclusion that approximate solutions to Eq. 2-42b are possible under three conditions:

1. If $C$ and $Q_a$ are such that the system lies well within the region between the two limiting lines of Fig. 2-10, Eq. 2-46* will give an accurate estimate of the pH.
2. If $Q_a$ is fairly large (greater than $10^{-6}$) and $C$ is relatively small (less than $0.1 \times Q_a$), the acid behaves essentially as a strong acid, and Eq. 2-10* is a reasonable estimate, subject to the limitations discussed earlier.
3. If $Q_a$ is small (less than $10^{-7}$) and $C$ is small enough that the product $CQ_a$ is not much larger than $10^{-13}$, the hydrogen ion produced by the autoprotolysis of water will play a significant role in establishing the pH of the solution. An approximate solution in this case will be developed after you have completed the following exercises:

EXERCISE 2.7:  Show that Eq. 2-46* follows from ignoring [X] in the mass balance (Eq. 2-40) and [OH⁻] in the proton balance (Eq. 2-41). Give a verbal, qualitative rationale for these two approximations.

EXERCISE 2.8:   What modifications of Eqs. 2-38–2-41 are necessary to produce Eq. 2-10*? Are these consistent with the description of the acid as "behaving essentially as a strong acid?"

An approximate equation for condition 3 may be obtained by realizing, from Fig. 2-10, that although we may ignore [X] in the mass balance for a very weak acid, we may not ignore [OH⁻] in the proton balance if the formal concentration of the acid is so low that [H⁺] may not be very much larger than [OH⁻]. If we solve Eq. 2-38 for [X] and insert the result in the proton balance, we obtain

$$[H^+] = \frac{Q_a[HX]}{[H^+]} + [OH^-] \qquad (2\text{-}47^*)$$

[OH⁻] is, without approximation and always, equal to $Q_w/[H^+]$; and if we ignore [X] in the mass balance and substitute $C$ for [HX], Eq. 2-47* becomes

$$[H^+] = \frac{Q_a C + Q_w}{[H^+]}$$
$$[H^+]^2 = (Q_a C + Q_w)$$
$$[H^+] = (Q_a C + Q_w)^{1/2} \qquad (2\text{-}48^*)$$

This equation is different from the simpler previous result, Eq. 2-45* only in the presence of $Q_w$ as an additive term representing the contribution of protons to the solution by the autoprotolysis of water, and which may be appreciable in certain extremes as indicated above. Examine the limit of Eq. 2-48* as either $Q_a$ or $C$ approach zero, and think of real chemical systems which fit these limiting conditions.

**Weak bases.**   An algebraic analysis of a $C\,F$ solution of a weak base in water begins with the same kinds of ideas as that for a weak acid in water. For the equilibrium system

$$B + H_2O = HB^+ + OH^- \qquad (2\text{-}49)$$

there are again four unknown concentrations, and the four independent relationships are

$$Q_b = \frac{[OH^-][HB]}{[B]}$$
$$Q_w = [H^+][OH^-]$$

mass balance:

$$C = [B] + [HB]$$

and proton balance relative to the reference level $H_2O + B$:

$$[H^+] + [HB] = [OH^-]$$

You should be familiar with this material; indeed, the mathematical situation is identical to that for a weak acid, with the difference that $OH^-$ now has the role played by $H^+$ in the weak-acid case, and $Q_b$ replaces $Q_a$. Clearly, we can approach this system most efficiently by writing the solutions obtained for the weak acid system, with $[OH^-]$ replacing $[H^+]$, and $Q_b$ replacing $Q_a$, obtaining numerical solutions for $[OH^-]$ (or for pOH), and then converting to $[H^+]$ or pH through $Q_w$.

> EXERCISE 2.9:   Write the weak-base analogs of Eqs. 2-39b, 2-40, 2-10*, 2-43*, 2-45* and 2-48*. Give the conditions under which the approximate equations apply. Sketch a graph analogous to Fig. 2-10, all for the case of a weak base dissolved in water. If you are skeptical of the validity of this process, derive the equations from the four independent relations above.

**Amphiprotic substances: algebraic treatment.**   In our graphical exploration of amphiprotic substances, we found a very straightforward and simple approximate solution, $pH = \frac{1}{2}(pQ_1 + pQ_2)$, and noted that this approximation was valid as long as the formal concentration of the amphiprotic substance was sufficiently large. Those of you who worked exercise 2.6 will recall that this approximation broke down at the not at all tiny concentration of 0.01 $F$ for $H_2PO_4^-$ and $HPO_4^{--}$. Clearly, we should have a good algebraic treatment for the fairly frequent cases like exercise 2.6 which will arise.

In a solution $C$ formal in the amphiprotic substance HA, there are five unknown concentrations: $[H_2A]$, $[HA]$, $[A]$, $[H^+]$, and $[OH^-]$, where $H_2A$ and A are, respectively, the conjugate acid and conjugate base of HA. Without any new ideas, we can find five independent equations:

$$Q_1 = \frac{[H^+][HA]}{[H_2A]} \qquad (2\text{-}50)$$

$$Q_2 = \frac{[H^+][A]}{[HA]} \qquad (2\text{-}51)$$

$$Q_w = [H^+][OH^-]$$

mass balance:

$$C = [H_2A] + [HA] + [A] \qquad (2\text{-}52)$$

and proton balance from the reference level HA + $H_2O$:

$$[H^+] + [H_2A] = [A] + [OH^-] \qquad (2\text{-}53)$$

An attempt to reduce these five to one equation in the one important unknown, $[H^+]$, results in a quartic equation which resists solution through the expedient of solving for $C$ (which appears to the first power) using assumed values of $[H^+]$. However, if we settle for an equation in two unknowns, $[HA]$ (which will certainly be close to $C$) and $[H^+]$, we will find that we can work our way toward a complete solution, picking up a valuable approximate equation along the way.

Beginning with the proton balance, we rearrange it and use the three equilibrium quotients:

$$[H^+] = [A] - [H_2A] + [OH^-]$$

$$= \left(\frac{Q_2[HA]}{[H^+]}\right) - \left(\frac{[HA][H^+]}{Q_1}\right) + \left(\frac{Q_w}{[H^+]}\right)$$

Multiplying both sides by $[H^+]$ and gathering terms in $[H^+]^2$, we find

$$[H^+]^2\left[1 + \left(\frac{[HA]}{Q_1}\right)\right] = Q_2[HA] + Q_w$$

or

$$[H^+] = \left(\frac{Q_2[HA] + Q_w}{1 + [HA]/Q_1}\right)^{1/2} \qquad (2\text{-}54)$$

We have not used the mass balance equation, which relates $[HA]$ to $C$. For ordinary values of $C$ (that is, larger than $10^{-6}$ $F$), $C$ and $[HA]$ will not be very different, and for a good approximation, we may substitute $C$ for $[HA]$ in Eq. 2-53:

$$[H^+] = \left(\frac{Q_2C + Q_w}{1 + C/Q_1}\right)^{1/2} \qquad (2\text{-}55^*)$$

EXERCISE 2.10: What happens to Eq. 2-55* when $C$ is large (that is, when $Q_2C \gg Q_w$, and $C/Q_1 \gg 1$)? Compare this result to the graphical solution.

We may carry on beyond Eq. 2-54 to an exact solution by solving the mass balance for $[HA]$ and substituting it in Eq. 2-54. However, we do not get a particularly tractable equation; it is much less workable than Eq. 2-42b.

EXERCISE 2.11: Try this substitution. Rearrange Eq. 2-54 to give $[HA]$ as a function of $[H^+]$, and see what happens when you substitute the approximate solution $[H^+]^2 = Q_1Q_2$ (Cf. exercise 2.10).

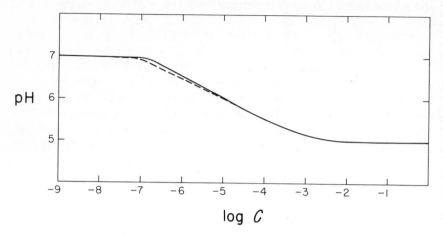

$$\log C$$

**Fig. 2-11.** pH *vs.* log $C$ for an amphiprotic substance whose conjugate acid has $pQ_1 = 3$, $pQ_2 = 7$. Solid line calculated from Eq. 2-54, dashed line from Eq. 2-55*.

Figure 2-11 is a graph of pH as a function of log $C$ for a model amphiprotic substance with $pQ_1 = 3$ and $pQ_2 = 7$. The lower line is a graph of Eq. 2-55*, and the upper line was calculated by using Eq. 2-54 to find $[H^+]$ as a function of assumed values of $[HA]$, and then using the mass balance in the form

$$C = \frac{[HA][H^+]}{Q_1} + [HA] + \frac{Q_2[HA]}{[H^+]} \qquad (2\text{-}56)$$

to calculate values of $C$ consistent with the assumed values of $[HA]$ and the calculated $[H^+]$. Thus, the upper line is an exact solution to the system. Note that Eq. 2-55* is correct to within 0.1 pH unit throughout the range of concentrations for this particular substance.

EXERCISE 2.12:   Derive Eq. 2-56 from the mass balance and the equilibrium quotients for this system. What is the significance of the high-$C$ limit of both lines in Fig. 2-11?

## Buffers

The degree of protonation and electrical charge of any Brønsted acid or base depends on the pH of the solution in which it finds itself. In turn, countless chemical and biological processes depend crucially on the degree of protonation of the acids and bases involved. For example, consider the fate of metabolically produced $CO_2$ in the body. The carbon dioxide cannot be allowed to accumulate at the site of metabolism, and in animals larger than insects it is not practical to provide direct gaseous exits to the external atmosphere. The ingenious solution is to hydrate the $CO_2$ to $H_2CO_3$, partially deprotonate this diprotic acid, and transport the resulting mixture of $H_2CO_3$, $HCO_3^-$, and $CO_3^{--}$ (along with dissolved but not hydrated $CO_2$) to the lungs via the bloodstream. In the lungs, the partial pressure of $CO_2$ in the gas phase is maintained at a low value by the familiar process called breathing, and in an effort to establish equilibrium between the solution and the gas phase, much of the $CO_2$ leaves the bloodstream and is exhaled. The decrease of $[CO_2]$ in the blood, of course, causes the equilibrium

$$H_2CO_3 = H_2O + CO_2$$

to shift to the right, and eventually most of the substance is removed from the body. Now consider the role of the pH of the bloodstream. If the pH is too high, too much $CO_2$ is retained in the body in the form of the nonvolatile ions $HCO_3^-$ and $CO_3^{--}$, since these predominate at a high pH; on the other hand, if the pH is too low, large concentrations of $CO_2$ and $H_2CO_3$ build up in the bloodstream, the solubility of $CO_2$ in the bloodstream is lowered (see Chapter 4), and the $CO_2$ is not efficiently carried away from the site of metabolism. (The latter condition induces a response known as "küssmaul breathing," as the body attempts to rid the bloodstream of its high $CO_2$ accumulation through drastic ventilation of the lungs. The name comes from the involuntary puckering of the mouth which accompanies the heavy breathing.)

As a second example, consider the indicator fluorescein, which is used to detect the end point of titrations of chloride ions with silver. In order to function, it is essential that the indicator, which is a weak acid, be in its negatively charged conjugate base form. If the pH of the titration mixture becomes too low, the titration fails because the indicator is in the neutral conjugate acid form and inert.

The response both of man and nature to situations in which the pH of a solution is important and subject to change through the introduction of acids or bases (as in the introduction of $H_2CO_3$ into the bloodstream) is to make the pH of the solution relatively insensitive to the addition of acids or bases. *Any solution that resists changes in its pH when an acid or base is added to it is called a buffer.*

There are two general classes of buffers. The most straightforward way of providing buffering is to have a relatively large amount of a strong acid or base already present in the solution. Then any added acid or base only slightly increases (or decreases) the amount of strong acid or base already present, and the pH change is small. For example, if we add one millimole of a strong acid to one liter of a 1 $M$ solution of a strong base, the $[OH^-]$

changes only by one part in 1000. The new $[OH^-]$ is related to the old by a factor of 0.999, and the pH by log 0.999, which is a change of $-0.0004$ units. This is excellent buffering action.

However, with a strong-acid or strong-base buffer one doesn't have much choice about the pH of the solution; it must be near either 0 or 14. As it happens, most reactions which need to be buffered occur at an intermediate pH. For example, the blood is buffered within a few hundredths of a unit of 7.35; departure from this value by as much as a few tenths of a unit produces severe physiological shock and/or death. Clearly a strong-base "buffer" of pH 7.35 would have to contain such a small concentration of $OH^-$ (roughly $1 \times 10^{-7}$ $F$ NaOH) that it would not be worthy of the name. The slightest influx of an acid or base would cause a profound change in pH.

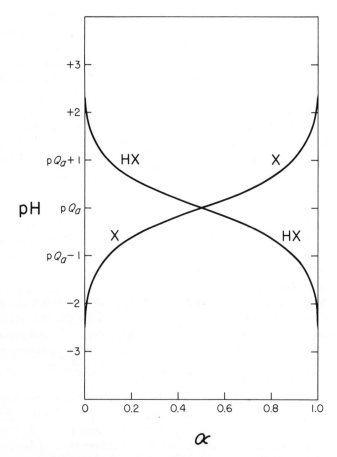

**Fig. 2-12.** pH *vs.* $\alpha_{HX}$ or $\alpha_X$ for a weak acid-conjugate base buffer. Compare Fig. 2-3.

To see how this dilemma is avoided, let us return to the linear distribution graph, Fig. 2-3, which is reproduced, rotated 90° in Fig. 2-12. Suppose we now consider the fractions of $\alpha_{HX}$ and $\alpha_X$ of a weak acid-conjugate base pair the independent variables in this graph, and the pH the dependent variable (the distinction is not fundamental). Look at either line and ask yourself when the pH is least sensitive to the fractional distribution. You should conclude that this occurs near the intersection of the two lines, where the pH is changing slowly as the fractions change rapidly. Now, if we had in a solution a mixture of a weak acid and its conjugate base, both at respectable concentrations, and attempted to change the pH by, say, adding a small quantity of a strong acid, the result would be

$$X + H^+ = HX$$

which would change the fractions $\alpha_X$ downward and $\alpha_{HX}$ upward somewhat, but the pH only relatively little, as long as we remain near the intersection of $\alpha_X$ and $\alpha_{HX}$ in Fig. 2-12. This will be true as long as both the weak acid and its conjugate base are present in comparable, fairly large concentrations or, to state it a third way, as long as the values of both $\alpha$'s remain near 0.5.

The beauty of this scheme is that we are now free to choose the pH of our buffer. True, for a given acid-base pair, $\alpha_X$ and $\alpha_{HX}$ will be near 0.5 when the pH is near the $pQ_a$ of the weak acid. However, we (and nature) are at liberty to choose the identity of the buffering acid-conjugate base pair. In the case of the blood, buffering constituents include $H_2PO_4^-$ and its conjugate base $HPO_4^{--}$.

EXERCISE 2.13:   What is the $Q_a$ of the conjugate acid in this pair? From Table 2.3, verify that the requirement that pH be near $pQ_a$ of the conjugate acid is realized for blood buffered by the phosphate system.

**Calculation of the pH of a buffer solution.**    To use weak acid-conjugate base buffers intelligently, and to prepare them to have a given pH, we need first to be able to calculate the pH of any given buffer. Suppose we know the formal concentrations $C_{HX}$ and $C_X$ of a weak acid and its conjugate base in a buffer solution. Our unknown quantities are [HX], [X], [H+], and [OH−]. The $Q_a$ of the weak acid may be rearranged to the following form, which promises an easy solution:

$$[H^+] = Q_a \frac{[HX]}{[X]} \qquad (2\text{-}57)$$

Certainly, the equilibrium molarities of HX and X are likely to approximate $C_{HX}$ and $C_X$, respectively. In fact, [HX] differs from $C_{HX}$ and [X] differs from $C_X$ only to the extent that they react with water according to the usual equilibria:

$$HX + H_2O = X + H_3O^+$$

$$X + H_2O = HX + OH^-$$

The situation is (superficially) complicated by the fact that the product of one reaction is the reactant of the other, and it is these equilibrium concentrations that we are trying to determine. However, without bothering to consider the above two equilibria separately, we realize that, since

$$[HX] + [X] = C_{HX} + C_X \tag{2-58}$$

$[X]$ will differ from $C_X$ by whatever amount $[HX]$ differs from $C_{HX}$, but in the opposite direction. This is seen by rearranging Eq. 2-57 to

$$C_{HX} - [HX] = -(C_X - [X]) \tag{2-59}$$

Suppose we call either side of Eq. 2-59 $D$; then Eq. 2-57 becomes

$$[H^+] = Q_a \frac{C_{HX} - D}{C_X + D}$$

We may relate $D$ to the other unknown concentrations by a proton balance beginning with the reference level $H_2O + HX$ (the substance X does not bring any loose protons to the solution). Since, as defined, $D$ represents a deficiency of protons relative to those added in the quantity $C_{HX}$, we add $D$ to the proton deficiency side of the balance:

$$[H^+] = [OH^-] + D \tag{2-60}$$

Note that there is no loss of generality in defining $D$ as we have; it could be a negative number and will be if the conjugate base X is more basic than the acid HX is acidic; that is, if $Q_a$ is less than $10^{-7}$.

We may now, having related $D$ to more prosaic variables, arrive at an exact equation for the $[H^+]$ in a weak acid-conjugate base buffer:

$$[H^+] = Q_a \frac{C_{HX} - ([H^+] - [OH^-])}{C_X + ([H^+] - [OH^-])} \tag{2-61}$$

This equation still contains two unknowns (assuming that we know the identity of the weak acid, and thus its $K_a$, and the formal concentrations $C_{HX}$ and $C_X$). However, the relation between $[H^+]$ and $[OH^-]$ in any solution is by now an elementary one.

We should now turn to solving Eq. 2-61. As it stands, Eq. 2-61 threatens trouble, since the unknown appears on both sides; we cannot make the corrections to $C_{HX}$ and $C_X$ representing the difference $D$ accurately unless we already know the $[H^+]$. If we expand Eq. 2-61 to put all the terms in $[H^+]$ on the left-hand side, we will find that it is a cubic equation. However, let us first see what can be done in the way of an approximate solution.

If our buffer is to be a good one, we have probably prepared it with respectable formal concentrations of the weak acid and its conjugate base; furthermore, the chances are that we are somewhere near the middle of the pH range, so that neither $[H^+]$ nor $[OH^-]$ is very large. In that case, $D$ will be quite small compared to $C_{HX}$ or $C_X$, and we come to an extremely useful and ubiquitous approximate equation:

$$[H^+] = Q_a \frac{C_{HX}}{C_X} \qquad (2\text{-}62^*)$$

This equation is known in medical and biological circles as *Henderson's Equation*, and we shall adopt this convenient label. Henderson's Equation is different from the exact truth only to the extent that $C_{HX}$ and $C_X$ are different, respectively, from $[HX]$ and $[X]$ by the difference $D$. If the formal concentrations are at least 0.1 $M$ (which is a fairly small value for an adequate buffer), $D$ will be no more than 1% of $C_{HX}$ or $C_X$ between the pH values 3 and 11.

EXERCISE 2.14:   Why?

That is, given omnipresent uncertainties in the precise value of $Q_a$ itself, Henderson's Equation is an excellent approximation for the majority of buffers encountered in practice.

However, there will arise the occasional situation in which the pH is near the high or low end of the scale, or the formal concentrations of the buffer constituents are rather low. If $D$ is appreciable, but not huge, compared to the formal concentrations, Henderson's Equation is still useful as a first approximation to obtain an estimate of $[H^+]$ and $[OH^-]$, from which $D$ may be calculated for a second approximation, and the exact $[H^+]$ may be deduced by successive approximations. To explore beyond this approach, we expand Eq. 2-61, after making the substitution $[OH^-] = Q_w/[H^+]$, to obtain

$$C_X[H^+] - C_{HX}Q_a = \frac{Q_a Q_w}{[H^+]} - [H^+]^2 - Q_a[H^+] + Q_w \qquad (2\text{-}63)$$

To assume various values of $C_X$ and $C_{HX}$ and solve this equation for $[H^+]$ as a test of Henderson's Equation would be possible, though tedious. We can obtain almost as much information about practical situations by

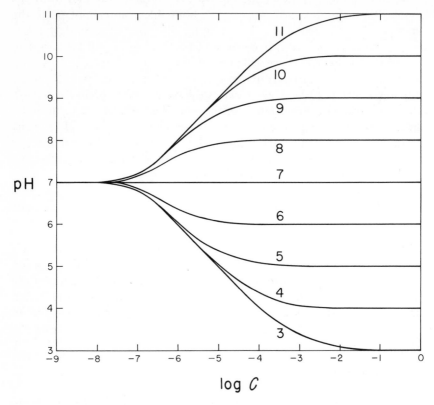

**Fig. 2-13.** pH of weak acid-conjugate base buffers with $C_{HX} = C_X = C$ vs. log $C$. The $pQ_a$ of the acid is indicated near its line.

assuming that $C_X$ and $C_{HX}$ are equal (as we have seen, they should be approximately equal for practical buffers), and solving the equation for $C$, the concentration of either constituent in this special case, as a function of assumed values of [H$^+$]. The results of this procedure are shown in Fig. 2-13 for weak acid-conjugate base pairs with a variety of $Q_a$'s. As in Fig. 2-10, a plot of pH vs. log $C$ was used for efficiency and aesthetics.

The horizontal straight portions at the right-hand (high-$C$) side of Fig. 2-13 are, of course, the Henderson's Equation predictions for buffers containing equal formal concentrations of the weak acid and its conjugate base: pH = $pQ_a$. The extent to which the lines representing the exact pH calculated from Eq. 2-61 deviate from Henderson's Equation varies in a regular way with the $pQ_a$ of the weak acid and the concentrations of the buffer constituents. Note that the closer the $pQ_a$ of the weak acid is to 7, the less is the deviation from Henderson's Equation at any value of $C$.

When Henderson's Equation predicts a pH of 7, it will be precise at any concentration, since the correction term $[H^+] - [OH^-]$ equals zero there.

The behavior of the pH of a buffer at the high-concentration end of Fig. 2-13 is the basis of the rough rule of thumb that the pH of a weak acid-conjugate base buffer is independent of dilution. This is certainly true to a first approximation, as long as the concentrations of the conjugate pair stay above roughly 10 times the $Q_a$ of the weak acid or the $Q_b$ of its conjugate base, whichever is larger.

EXERCISE 2.15: Verify the above rule roughly by inspection of Fig. 2-13. How does it follow from Henderson's Equation?

EXERCISE 2.16: The lines for conjugate pairs whose $pQ_a$ is less than 6 or greater than 8 contain linear portions. The equation of the linear portion for acids with $pQ_a$ less than 6 is pH $= -\log C$; that for acids with $pQ_a$ greater than 8 is pH $= 14 + \log C$, or pOH $= -\log C$. Why do you suppose this should be? (Where have we seen those equations before?)

**Preparation of buffers.** Because it is a common practical problem, this is a good place to discuss the preparation of a buffer with a particular pH.

1. Because we generally want to prepare buffers which will have adequate capacity to resist changes in their pH, let us assume that the concentration of the weak acid and its conjugate base will be large enough for Henderson's Equation to be used with confidence. Since, as we have seen, buffers work by allowing a small part of one member of the conjugate pair to be converted to the other member, with minimal resulting change in pH, the larger the concentrations of the two species, the better is the buffer. Other considerations (such as solubility, or a desire to limit the total concentration of salts in the solution) may put an upper limit on $C_{HX}$ and $C_X$, but we will assume that they are each at least 0.05 $M$. From Fig. 2-13, this concentration allows the use of Henderson's Equation with good precision for $pQ_a$'s between 3 and 11.

2. Bearing in mind that the requirement that both the weak acid and its conjugate base should have appreciable concentrations means that $pQ_a$ of the acid should be close to the desired pH, we consult a $pQ_a$ table of weak acids, not neglecting polyprotic acids. A $pQ_b$ listing of weak bases will also be a source of eligible acids; mentally subtract the listed $pQ_b$ from 14 to arrive at the $pQ_a$ of the conjugate acid. At this point we choose acids for the various secondary features we may desire. For example, in a solution that will be in contact with living organisms, we would want to avoid toxic substances; or in an experiment involving high temperatures or contact with nonpolar solvents, we would want to avoid electrically neutral buffer components, which might volatilize or enter a nonaqueous phase. In this case, a member of a

polyprotic acid's family such as $HPO_4^{--}$-$PO_4^{3-}$, etc., would be a good choice.

3. Unless we are quite lucky, we will not find an acid whose $pQ_a$ is precisely equal to the pH we want our buffer to have. It remains, having chosen the conjugate pair, to calculate the exact ratio of acid to conjugate base required to give the desired pH. We rearrange Henderson's Equation as follows:

$$\frac{C_{HX}}{C_X} = \frac{[H^+]}{Q_a}$$

Since both factors on the right-hand side are known, the calculation of the proper ratio is straightforward.

4. If there are no particular requirements on the values of $C_{HX}$ or $C_X$, we choose any convenient value, probably several tenths $F$, for the member which will have the smaller concentration, and calculate the concentration of the other which is in the proper ratio. This procedure insures that neither constituent of the buffer will have a concentration so low as to be ineffective in buffering.

Let us illustrate this procedure with a typical example. Suppose we are interested in measuring the respiration rate of some bacterium immersed in a solution of pH 9.0. Consulting a table of acid dissociation constants, we readily find at least two possibilities: HCN, with a $pK_a$ of 9.32, and $NH_4^+$, with a $pK_a$ of 9.26. There are three reasons for discarding HCN, besides the fact that its $pK_a$ happens to lie a little farther from the desired pH than that of $NH_4^+$: first, it is somewhat volatile; second, it is dangerously toxic to the experimenter; and third, the metabolic rate of the bacterium would be likely to be seriously altered in its presence. Settling on $NH_4^+$, we calculate the ratio $[H^+]/Q_a$ which is equal to $C_{HX}/C_X$:[2]

$$[H^+] = 1.0 \times 10^{-9}$$

$$Q_a = 5.5 \times 10^{-10}$$

$$\frac{[H^+]}{Q_a} = \frac{C_{NH_4^+}}{C_{NH_3}} = 1.8$$

Any concentrations larger than $10^{-3}$ $F$ would, from Fig. 2-13, be high enough for the Henderson's-Equation calculation to be accurate; for a good buffer, we want relatively high concentrations, so suppose we decide on 0.1 $M$ $NH_3$ and 0.18 $M$ $NH_4^+$, which we can obtain by making the solution 0.18 $F$ in $NH_4Cl$. If we have reason to suspect that the solution will be

---

[2]If the precision of our calculation warrants it, we should correct $K_a$ for salt effects, at the ionic strength of our proposed buffer, before making this calculation. See Appendix 2.

subjected to large influxes of acid or base, we can increase the *buffer capacity* (defined as the ratio of the number of moles of added $H^+$ or $OH^-$ to the pH change which they cause) by increasing the concentrations of the buffer constituents, e.g., to 0.5 $M$ $NH_3$-0.9 $F$ $NH_4Cl$, or 1 $M$ $NH_3$-1.8 $F$ $NH_4Cl$.

## Acid-Base Equilibria in Non-Aqueous Solvents

As we have remarked, most of the important naturally occurring solutions are aqueous ones — solutions whose major component is water. These solutions include all known forms of life, the oceans, rivers, and (partly) the soil, and much of the subject matter of inorganic chemistry. However, water has certain limitations as a solvent which have led to the investigation of other solvents for ionic equilibria, especially proton-transfer reactions. Of course, organic chemists, who deal mainly with non-polar molecules, have used nonpolar or slightly polar solvents such as hydrocarbons, halocarbons, ethers, and the like for decades. The systematic exploration of polar, nonaqueous liquids as solvents for ionic systems, however, is a relatively recent development. We shall see that much of the arithmetic we have developed for dealing with aqueous equilibria is directly transferrable to other solvents; but each solvent has its own person-ality dependent on factors we have taken for granted in water.

The most striking limitation of water as a solvent is its inability to dissolve nonpolar solutes. We regularly prepare fairly concentrated solutions of sugar or salt in water, but no amount of shaking will make the oil mix with the vinegar in a salad dressing. Equilibrium studies on organic acids or bases with large nonpolar parts may be impossible to carry out in water, just because the solubility of such substances in water is gener-ally very slight. Changing to a less polar solvent such as methanol, pure ("glacial") acetic acid, dimethyl sulfoxide, or acetone may solve this problem. The trick is to find a liquid non-polar enough to tolerate the nonpolar portion of the solute molecule, but still polar enough to permit the presence of ions, either by having a moderately large dielectric constant or by interacting strongly with the solute through coulombic effects such as hydrogen bonding. A small but growing group of solvents has been found to be useful. A representative sampling of these will be found in Table 2.4.

In common with other solvents water possesses a more subtle limitation: the strengths of the acids or bases compatible with it. In some situations, most notably in acid-base titrations, one is interested in preparing a solution of the strongest proton acceptor or donor possible. As we have seen when discussing the pH of the solutions of strong acids and bases, any base stronger than the hydroxide ion will react with water to produce an equal quantity of $OH^-$; likewise, any acid stronger than the hydronium ion will

TABLE 2.4.  SOME POLAR NONAQUEOUS SOLVENTS

| SOLVENT | DIELECTRIC CONSTANT | $pK_s$ |
|---------|---------------------|--------|
| $H_2SO_4$ | 110 (20° C) | 3.62 |
| Acetic acid | 6.13 | 14.45 |
| Formic acid | 58.5 | 6.2 |
| $NH_3$ | 22 ($-33°$ C) | 28 |
| Ethylenediamine | 12.9 | 15.3 |
| Methanol | 31.5 | 16.7 |
| Ethanol | 24.2 | 19.1 |
| Formamide, $H-C{\overset{O}{\underset{NH_2}{\big<}}}$ | 109 | — |
| Nitromethane, $H_3CNO_2$ | 35.9 | — |
| Acetone, $H_3C-\underset{\underset{O}{\|}}{C}-CH_3$ | 20.4 | — |
| $SO_2$ | 17.27 ($-16.5°C$) | — |

simply protonate water molecules to produce an equal quantity of $H_3O^+$. Consequently $H_3O^+$ is the strongest acid, and $OH^-$ the strongest base, that can exist in water solutions. If we want, for example, a solution of a stronger base, we must switch to a solvent whose conjugate base is a better proton acceptor than $OH^-$. Ethylenediamine ($H_2N(CH_2)_2NH_2$), whose conjugate base is $H_2N(CH_2)_2NH^-$, is such a solvent. Conversely, for a stronger acid than $H_3O^+$, we seek a solvent such as glacial acetic acid, whose conjugate acid ($H_3CC(OH)_2^+$) is stronger than the hydronium ion.

If a solvent is both appreciably acid and appreciably basic (as water is), it will have a measurable autoprotolysis constant. Such solvents are called protolytic. Examples are methanol:

$$2H_3COH = H_3CO^- + H_3COH_2^+ \qquad K_s = 2 \times 10^{-17}$$

liquid $NH_3$:

$$2NH_3 = NH_2^- + NH_4^+ \qquad K_s = 10^{-28}$$

and glacial acetic acid:

$$2H_3CCOOH = H_3CCOO^- + H_3CC(OH)_2^+ \qquad K_s = 3.6 \times 10^{-15}$$

$K_s$ in each of the above equations is the analog of $K_w$ for water; the equilibrium constant for autoprotolysis with mole fraction used as the concentration unit for the solvent. Not all solvents have autoprotolysis constants smaller than that of water, as may be seen in Table 2.4.

Likewise, there is no reason not to use the same definitions of $Q_a$ and $Q_b$ for solute acids or bases:

$$HX + S \text{ (any solvent)} = X + HS^+ \qquad Q_a = \frac{[HS^+][X]}{[HX]}$$

$$X + HS \text{ (any protolytic solvent)} = HX + S^- \qquad Q_b = \frac{[S^-][HX]}{[X]}$$

The numerical value of any $K_a$ clearly depends not only on the inherent acidity of the acid, but on the basicity of the solvent. A really basic solvent can snatch a proton from even a reluctant acid, giving it a large $K_a$. As a general rule, acids look strongest in basic solvents, and bases look strongest in acidic solvents. This rule is subject to some modification depending on the dielectric constant of the solvent, which we will discuss later.

If the solvent is basic enough, a series of intrinsically different acids will look perfectly strong in it; the $K_a$'s will all be so large that the amount of the undissociated acid HX will be undetectable regardless of their inherently different strengths. This is the case for HCl and $HClO_4$ in water, both appearing to be (and being) categorically strong; however, they are both incompletely dissociated in glacial acetic acid, and their strengths turn out to lie in the order $HClO_4 > HCl$. The inherent differences in the acid strengths are said to be *leveled* by the water. Similarly, differences in basicity are leveled if the solutes are stronger bases than the solvent's conjugate base. A solvent that does not level two acids or bases is said to *differentiate* them.

Another important factor which governs the numerical value of $K_a$ should be mentioned here. If the dissociation of an acid requires the separation of charge (as it does for all electrically neutral or negative acids), the $K_a$ of that acid will be sensitive to the dielectric constant of the solvent. The proton may be transferred to the solvent, but unless coulombic forces are sufficiently weakened by the solvent's dielectric constant, the resulting ions will not separate, but will remain together as an *ion pair*, that is, as two ions held together tightly by coulombic attraction. For example, consider HCl in glacial acetic acid (whose dielectric constant is about 6, less than that of water by a factor of 13):

$$HCl + H_3CCOOH = H_3CC(OH)_2^+Cl^- = H_3CC(OH)_2^+ + Cl^-$$

Whereas the intrinsic acidity of HCl relative to acetic acid may be said to be compared in the first equilibrium, the $K_a$, by definition, relates the products and reactants of the over-all two-step process, and thus describes not only the strength of HCl as an acid, but the willingness of the ions formed to part.

As we might expect, acids that are positively charged to begin with are not so sensitive to the dielectric constant of the solvent, because when they transfer a proton to a neutral solvent molecule, no separation of charge occurs:

$$HX^+ + S = HS^+ + X$$

Thus, the $K_a$ of acetic acid decreases by a factor of more than $10^5$ when the solvent is changed from water to ethanol, whereas that of anilinium ion ($C_6H_5NH_3^+$) decreases by a factor of about 10 when the same two solvents are compared. Clearly, most of the effect on the $K_a$ of acetic acid is in the change of dielectric constant (that of water is 79; ethanol, 25), rather than in the difference in the basicity of the solvents.

## Suggestions for Further Reading:

Bell, R. P., *The Proton In Chemistry*. Ithaca, N. Y.: Cornell University Press, 1959. This excellent short treatment has the generality promised by the title and includes a good discussion of structural influences on the $K_a$.

Butler, James N., *Ionic Equilibrium: A Mathematical Approach*. Reading, Mass.: Addison-Wesley Publishing Co., 1964. Chapters 4, 5, and 7 deal with acid-base equilibria in Butler's elegant and clear style. Graphical approaches are used throughout.

Fleck, George M., *Equilibria in Solution*. New York: Holt, Rinehart and Winston, Inc., 1966. Chapters 3 through 6 explore in detail acid-base equilibria in water and other solvents, especially with regard to titrations.

Kolthoff, I. M., P. J. Elving, and E. B. Sandell, eds., *Treatise on Analytical Chemistry*. New York: Interscience Encyclopedia, 1959. Part I, Volume 1, Chapter 8 of this compendium is a complete discussion of the graphical presentation of equilibrium data by L. G. Sillén. Chapters 10 through 13, by R. G. Bates, I. M. Kolthoff, and S. Bruckenstein, cover acid-base equilibria.

Sisler, H. H., *Chemistry in Non-Aqueous Solvents*. New York: Reinhold Publishing Corp., 1961. A brief treatment of the principles and particular properties of selected solvents.

Waddington, T. C., ed., *Non-Aqueous Solvent Systems*. New York: Academic Press, 1965. More complete discussions, solvent by solvent, each written by specialists on that solvent. A very valuable book to consult if you are planning to use a particular solvent for an ionic system.

## Problems

2.1.  Write a proton balance, a charge balance, and a mass balance for a solution of sodium hydrogen tartrate (NaHT) in $H_2O$.

2.2.  Write proton balances for the following solutions:

A. $HOCl + H_2O$

B. Methylamine $(CH_3NH_2) + H_2O$

C. $H_2SO_4$ + ethylenediamine $(H_2NCH_2CH_2NH_2)$

D. $NaNH_2 + CH_3OH$

E. $H_2C_2O_4 + H_2O$

F. $NaHC_2O_4 + H_2O$

G. $Na_2C_2O_4 + H_2O$

2.3. Calculate the pH of the following solutions. Do at least two problems both graphically and algebraically, and the rest as seems best to you. The last three are not at equilibrium.

A. 0.01 $F$ acetic acid

B. $1 \times 10^{-6}$ $F$ HCl

C. $1 \times 10^{-6}$ $F$ acetic acid

D. $1 \times 10^{-4}$ $F$ phenol

E. 0.05 $F$ $Na_2S_2O_3$

F. 0.01 $F$ ethylenediamine

G. 0.1 $F$ $NaHCO_3$

H. 0.1 $F$ $NH_3$

I. 0.1 $F$ $NH_3$ + 0.03 $F$ $NH_4Cl$

J. 0.1 $F$ $NH_3$ + 0.05 $F$ $HNO_3$

K. 0.1 $F$ NaOH + 0.15 $F$ $H_3PO_4$

L. $1 \times 10^{-4}$ $F$ NaOH + $1 \times 10^{-4}$ $F$ phthalic acid

2.4. Draw a logarithmic distribution diagram for 0.1 $F$ ammonium acetate. What is the pH of this solution? To what system discussed in this chapter is this similar?

2.5. A solution was prepared to contain 1.0 $F$ chloroacetic acid (HClA) and 0.5 $F$ NaClA, both of which are colorless. An indicator, $2.97 \times 10^{-3}$ $F$ m-nitroaniline (MNA), was present, and by a careful spectrophotometric measurement of the solution's color, it was found that the yellow basic form MNA was $1.76 \times 10^{-3}$ $M$. The p$Q_a$ of HMNA$^+$ in this solution was found to be 2.697. Calculate the [HMNA$^+$], [H$^+$], and the $Q_a$ and p$Q_a$ of HClA in this solution.

2.6. How valid would Henderson's Equation be for a chloroacetic acid-sodium chloroacetate buffer of the composition given above?

2.7. Suppose that an indicator whose acid form is red and whose basic form is yellow is used in an acid-base titration. To an "average observer" the solution looks completely red if the ratio $R$ for this indicator is greater than 5, and looks completely yellow if $1/R$ is greater than 8. If the $Q_a$ of this indicator is $3 \times 10^{-6}$, over what range of pH will the color appear to change from red to yellow?

2.8. Amino acids feature an amine group ($-NH_2$) and a carboxylic acid group on the same molecule. The biologically important $\alpha$-amino acids, from which proteins are assembled, have the general structure

The amine group is on the $\alpha$ carbon, i.e., the one nearest to the carboxyl group. R represents the remainder of the molecule, and may be simple, like a methyl

group, or quite complex; in particular, it may contain other acidic or basic groups. In water solutions, the proton of the carboxyl group is found to transfer to the amine group. The resulting species, having both positive and negative charges on the same molecule, is called a *zwitterion;* its structure is

$$
\begin{array}{c}
H \\
| \\
R-C-COO^- \\
| \\
_+NH_3
\end{array}
$$

The zwitterion may be considered the conjugate base of a diprotic acid. Construct a logarithmic distribution graph for the amino acid $\alpha$-alanine, and determine the pH of 0.1 $F$ $\alpha$-alanine.

2.9.  Aspartic acid,

$$
\begin{array}{c}
H \\
| \\
HOOC-CH_2-C-COOH \\
| \\
NH_2
\end{array}
$$

is an amino acid with an extra carboxyl group, so that the zwitterion is the conjugate base of a triprotic acid. Assuming that, as in other amino acids, the ammonium group is the last to deprotonate, give the structure and charge of the four conjugate forms.

2.10.  Electrophoresis is a technique by which mixtures of amino acids are separated into their components on the basis of their migration in an electric field; positive species migrate toward the negative pole, and vice versa. How is this behavior related to the structure of the amino acid and the pH of the solution? At what pH might glycine and aspartic acid be separated by electrophoresis?

2.11.  The pH at which an amino acid does not migrate in an electric field is called the *isoelectric point*. What is the predominant form of alanine at the isoelectric point? Of aspartic acid? What are the isoelectric pH's of these two acids?

2.12.  For a solution of a diprotic acid $H_2X$ in water, show that $[X^{-2}]$, and thus the contribution of the second dissociation to the $[H^+]$, is approximately equal to $Q_2$. (Hint: substitute the approximate proton balance into $Q_2$.)

# 3 Coordination Equilibria

## Introduction: Reactions and Definitions

When a proton leaves one molecule or ion and settles on another, we call the transaction a Brønsted acid-base reaction. From a wider perspective, we might be inclined to treat a proton as just one example of a small, electron-deficient species drawn coulombically to electron-rich sites. Its unique importance stems from the fact that it is a necessary component of that most important of all solvents, water. If we take a liberal view, we may include as *acids* all electron-poor species that form more or less covalent bonds with an otherwise nonbonding pair of electrons on an electron-rich base. This classification is due to G. N. Lewis, and reactions in which a *coordinate* bond is formed through the *sharing* of an electron pair between a donor base and acceptor acid are called *Lewis acid-base reactions*. Alternatively, such reactions are called complexation reactions since from two species capable of independent existence, a complex compound (or simply a *complex*) is formed. Examples are found in practically every branch of chemistry; a modest collection of Lewis acid-base reactions is given in Table 3.1.

Most of the Lewis acids found in water solutions are metal ions, all of which, being positive, are to some extent electron deficient. The fascination of metal complexes has helped to revolutionize the once moribund science of inorganic chemistry, and has, as a possibly less happy result, spawned some rather specialized terminology as follows:

*Ligand:* An atom, molecule, or ion containing one or more Lewis basic sites. Examples are $H_2N(CH_2)_2NH_2$, $H_2O$, and $(CH_3)_3P$.

*-dentate:* A suffix indicating the number of sites at which a ligand is bonded to a metal ion. $Cl^-$ is monodentate, $H_2\ddot{N}(CH_2)_2\ddot{N}H_2$ bidentate, etc.

*Chelon:* A polydentate ligand,[1] usually containing O and N atoms as the basic sites.

*Chelate:* A metal-chelon complex.

*Coordination number:* The number of bonds formed by each metal to all of its ligands. Thus the coordination number of cobalt in the compound $Co(H_2O)_5Cl^{++}$ is six: there are five bonds to water molecules and one to the chloride ion.

TABLE 3.1. SOME LEWIS ACID-BASE REACTIONS

| LEWIS ACID | LEWIS BASE | PRODUCT | COMMENTS |
|---|---|---|---|
| $BF_3$ | $NH_3$ | $F_3BNH_3$ | |
| $H^+$ | $NH_3$ | $NH_4^+$ | Note that in this system the proton itself is an acid, and that any Brønsted base is also a Lewis base. |
| $Fe^{+3}$ | $6H_2O$ | $Fe(H_2O)_6^{3+}$ | Most metal ions exist in water solutions as *aquo* complexes if no other Lewis base is present. |
| $Co(H_2O)_6^{3+}$ | $6NH_3$ | $Co(NH_3)_6^{3+}$ | A six-step ligand exchange reaction. |
| $Ag^+$ | $Cl^-$ | $AgCl$ | The product of this reaction is a molecular 1-to-1 adduct, not the ionic crystal. |
| Hemoglobin | $O_2$ | Oxyhemoglobin | The iron atom in hemoglobin binds oxygen for transport to a metabolic site. |

Since the maximum possible coordination number of most metal ions is larger than the number of basic sites of most ligands, the possibility arises of several-step reactions, analogous to those of polyprotic Brønsted acids and bases. As in Brønsted equilibria, stepwise equilibrium constants are the most frequently used. However, there seems to be a preference for *association* or *stability* constants rather than dissociation constants. For example, in the $Cu^{++} - NH_3$ set of complexes,

$$Cu^{++} + NH_3 = CuNH_3^{++} \qquad K_1 = \frac{(CuNH_3^{++})}{(Cu^{++})(NH_3)}$$

$$CuNH_3^{++} + NH_3 = Cu(NH_3)_2^{++} \qquad K_2 = \frac{(Cu(NH_3)_2^{++})}{(CuNH_3^{++})(NH_3)}$$

etc.

[1]When one unscrambles the Latin and Greek roots of these terms, the result is the mixed metaphor: a string with many teeth is a claw.

Metal ions in water solutions are Lewis acids literally swamped in a very decent Lewis base, the $H_2O$ molecule. The unadorned symbol $Cu^{++}$ therefore really means $Cu^{++}$ coordinated to several water molecules, and when any other ligand bonds to the $Cu^{++}$, one or more water molecules must leave. Since it can be tedious to write the water part of these *aquo* complexes, the coordinated water molecules are usually omitted and understood to be present unless the maximum possible coordination number of the metal ion (which depends partly on its size and partly on its electronic structure) is occupied by other ligands.

The number of water molecules bonded to each metal ion (the *hydration number*) is not always easy to determine and depends partly on what is meant by "bonded." Most transition metals bind six water molecules by essentially covalent bonds, but since a hexaquo complex ion is still electrically positive, neighboring water molecules will still be attracted, and the effective hydration number may be much larger than six. For weakly acidic ions like $Na^+$, $Ca^{++}$, etc., the interaction with water molecules is even less well-defined than that of transition metals, and the hydration number may depend on the concentration of the metal ion in the solution.

As we have seen in Chapter 2, some metal ions interact with water so strongly as to produce Brønsted acids (e.g., $Al(OH_2)_n^{+3}$, which has a $pK_a$ of 5). These ions are present as the "simple" undissociated aquo ions only at low pH.

## Factors that Influence the Stability Constant

If we now consider the question *why* one stability constant is larger or smaller than another, any generalizations we make must cover much more ground than those we made for Brønsted equilibria. In the latter class of reactions, the question was simply one of relative abilities of bases to bind a given particle — the proton. Now we must compare not only bases, but also the reactivity of these bases toward a variety of acids. As a first guess, we might feel that the smaller and more like a proton a given acid is (since a proton represents the ultimate in charge-to-radius ratio) the more tightly it would be bound by any base. A high charge and small radius should allow an acid to get close to a base and interact strongly with the latter's abundance of electrons. To some extent this is a valid criterion; for example, the stability constants $K_1$ of the hydroxide complexes of the following metals increase in order: $Na^+ < Ca^{++} < Mg^{++} < Al^{3+}$. On the other hand, toward the base $S_2O_3^{--}$, the acidity of the following ions increases in order: $Fe^{3+} < Cd^{++} < Ag^+$; and among the alkaline earth metals, with a constant charge of $+2$, $K_1$ for complexation with $S_2O_3^{--}$ increases in the order: $Ca^{++} < Sr^{++} < Ba^{++}$, even though the radii of these ions increase in the same order. Clearly, $S_2O_3^{--}$ and $OH^-$ have different criteria for attractive acids.

A recent semiempirical correlation classifies acids and bases as "soft" or "hard," according to the ease with which their electron clouds may be distorted by the electric field of a nearby atom or ion. The terms are intended to be descriptive in a rough way: soft bases have loosely held lone pairs in large, easily distorted orbitals, whereas hard ones draw their lone pairs in tightly; soft acids are characterized by fairly abundant electrons in outer orbitals which are screened from the nucleus, whereas hard acids have few if any $d$ electrons, and high charge-to-radius ratios, resulting in tightly held, closely packed electron shells. For example, most of the transition metals in low oxidation states toward the right-hand side of the transition series are either soft or "intermediate" acids. Examples of hard acids are the alkali metal ions, the alkaline earth metal ions, and $Al^{3+}$, all of which have electronic configurations characteristic of a rare gas and a nuclear charge several units higher than that rare gas. The proton is the pre-eminent example of a hard acid, because it has no electrons at all, and a huge charge-to-radius ratio.

The point of this correlation is that hard acids tend to form their most stable complexes with hard bases, and soft acids with soft bases. Of course many hard-soft combinations do result in stable complexes, but these tend to be less stable than comparable hard-hard or soft-soft complexes. This rule is exemplified in the data given above for $OH^-$ and $S_2O_3^{--}$ complexes. The oxygen atom is very electronegative and keeps its electrons tightly pulled in. As a result, most Lewis bases with oxygen atom donor sites tend to react as hard bases; the hydroxide ion is a good example. Thus, the order of increasing stability in the series of complexes $NaOH$, $CaOH^+$, $MgOH^+$, $AlOH^{++}$ is also the order of increasing charge-to-radius ratio and of increasing hardness of the Lewis acid. On the other hand, toward the soft base $S_2O_3^{--}$, which binds metal ions through a sulfur atom ($O_3SS{:}M$), the acidity in both the transition metal and alkaline earth metal series was in order of decreasing charge-to-radius ratio or decreasing hardness.

Various theories may be advanced to account for the above empirical correlations (which may be extended to other acids than metal ions, to other solvents, and to kinetic as well as to equilibrium considerations). One factor is surely the simple coulombic one; if no more special forms of bonding are possible, the closer together the acid and base can get, the stronger are the coulombic forces between them. This seems to account for most of the hard-hard interactions, because tight little ions can approach each other closely. This is simply an extension of the argument given to account for the decrease in proton basicity in the series $F^- > Cl^- > Br^- > I^-$. Soft-soft interactions appear to involve more complex bonding. It is noteworthy that most metal ions which are soft acids possess filled or nearly filled $d$ subshells in their outer shells. These may be in a position to donate electron density to unfilled, pi-symmetry orbitals which most soft ligands have (Fig. 3-1). Such "back" or "push-pull" bonding has been invoked to account for many unusually stable bonds, because it allows ligands to

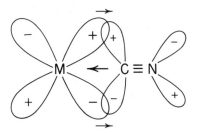

**Fig. 3-1.** Metal-ligand double bonding in a soft acid-soft base complex. The leftward arrow represents an "ordinary" sigma bond formed with electrons donated by the ligand (in this example, a carbon lone pair on a $CN^-$ ion); the rightward arrows represent a pi bond formed by donation of electrons from the metal into an orbital formed from a metal $d$ orbital and a vacant antibonding molecular orbital of the ligand.

coordinate with a metal without an embarrassing pileup of negative charge on the electropositive metal. A complete explanation of the stabilities of metal-ligand complexes is beyond the scope of this book and, in fact, beyond the scope of any chemist at present.

## Distribution of Species in Multistep Lewis Equilibria

Given values of the stepwise stability constants and the equilibrium ligand concentration, the equilibrium

$$M + L = ML$$
$$+$$
$$L = ML_2$$
$$+$$
$$L = ML_3 \cdots \text{etc.}$$

may be mapped out in the same way that a polyprotic Brønsted equilibrium is. From mass balance,

$$C_M = [M] + [ML] + [ML_2] + [ML_3] + \cdots \qquad (3\text{-}1)$$

The fraction $\alpha_0$ of the metal present as the uncomplexed M may be derived by dividing Eq. 3-1 by [M] to obtain

$$\frac{C_M}{[M]} = \frac{1}{\alpha_0} = 1 + \frac{[ML]}{[M]} + \frac{[ML_2]}{[M]} + \frac{[ML_3]}{[M]} + \cdots$$

The second term on the right-hand side of this series is easily obtained from $Q_1$:

$$Q_1 = \frac{[ML]}{[M][L]}$$

$$\frac{[ML]}{[M]} = Q_1[L] \tag{3-2}$$

and the third term from the product $Q_1 Q_2$:

$$Q_1 Q_2 = \frac{[ML][ML_2]}{[M][ML][L]^2}$$

$$\frac{[ML_2]}{[M]} = Q_1 Q_2[L]^2 \tag{3-3}$$

EXERCISE 3.1:   Write the term for $[ML_3]/[M]$ by induction from 3-2 and 3-3. Confirm your guess by formally deriving it, and write a general expression for $[ML_n]/[M]$.

We now have

$$\frac{1}{\alpha_0} = 1 + Q_1[L] + Q_1 Q_2[L]^2 + \cdots \tag{3-4a}$$

or

$$\alpha_0 = \frac{1}{1 + Q_1[L] + Q_1 Q_2[L]^2 + \cdots} \tag{3-4b}$$

Once we have obtained $\alpha_0$ as a function of the ligand concentration, the remaining $\alpha$'s — the fractions of $ML$, $ML_2$, etc. — do not present such a formidable array of algebra. For example, consider $\alpha_1$, the fraction $[ML]/C_M$.

$$\frac{[ML]}{C_M} = \frac{[M]}{C_M}\frac{[ML]}{[M]} \tag{3-5}$$

which is to say

$$\alpha_1 = \alpha_0 \frac{[ML]}{[M]} \tag{3-6}$$

The ratio $[ML]/[M]$ is easily obtained from $Q_1$ as before:

$$\frac{[ML]}{[M]} = Q_1[L] \tag{3-7}$$

Therefore,

$$\alpha_1 = \alpha_0 Q_1[L] \qquad (3\text{-}8)$$

EXERCISE 3.2:   Derive expressions for $\alpha_2$ and $\alpha_3$ analogous to Eq. 3-8, and write a general equation for $\alpha_n$ as a function of $\alpha_0$ and [L].

## Distribution Graphs

When all the $\alpha$'s for a given system have been calculated, they may be graphed, and problems solved graphically, in the same manner discussed for Brønsted acid-base systems in Chapter 2. The mathematical implications of the reaction of one species with another in a series of steps are the same regardless of the chemical names for the species. Figure 3-2 is a logarithmic distribution graph for the $Ag^+ - S_2O_3^-$ system.

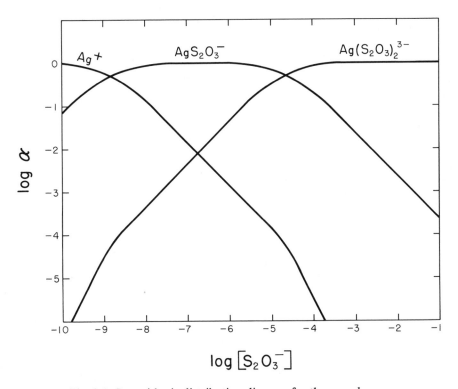

**Fig. 3-2.** Logarithmic distribution diagram for the complexes formed by $Ag^+$ and $S_2O_3^-$ in aqueous solutions at 25° C.

EXERCISE 3.3: Show algebraically that the system points for this graph are at $\log [L] = -\log Q_1$ and $\log [L] = -\log Q_2$.

The unique advantage of a logarithmic distribution graph is its linearity, which allows any particular graph to be drawn quickly once values of the stepwise formation constants are known. When we were discussing such graphs quantitatively in Chapter 2, we found that this linearity is only a valid approximation as long as the log of the ligand concentration (in the discussion of Chapter 2, the pH) is not close to any of the log Q's. In Fig. 3-2, as in all other such graphs, the lines are obviously nonlinear in the neighborhood of $-\log [S_2O_3^{--}] = \log Q_1$ and $-\log [S_2O_3^{--}] = \log Q_2$. For many (indeed most) multistep Lewis acid-base equilibria, the successive stability constants are much closer together than are those for the $Ag^+$-$S_2O_3^{--}$ system. As one example among many, the logarithms of the first five stability constants of the $Cu^{++}$-$NH_3$ system are 4.15, 3.50, 2.89, 2.13, and $-0.5$. When the stepwise constants are spaced this closely, one might as well graph $\alpha$, rather than log $\alpha$, especially since several species

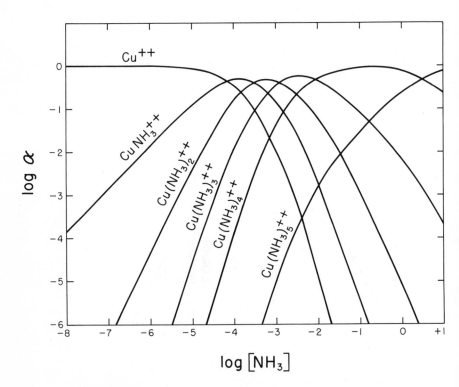

**Fig. 3-3.** Logarithmic distribution diagram for the $Cu^{++}$-$NH_3$ complexes.

may well have values of $\alpha$ between 0.1 and 0.9 simultaneously, and it is just within this region that a logarithmic graph is least helpful. In a form used with particular frequency in Lewis acid-base equilibria, the fractions of the various species are graphed as fractions of a vertical distance which represents unity, rather than allowing them to overlap as in Fig. 2-3. Let us illustrate with the $Cu^{++}$-$NH_3$ system:

Assuming values of log [$NH_3$], we may calculate $\alpha_0$, $\alpha_1$, $\alpha_2$, etc., from Eqs. 3-4, 3-8, and your results from Exercise 3.2. A logarithmic distribution graph based on this calculation is shown in Fig. 3-3. Note that over a range of about 100-fold in [$NH_3$], from $10^{-2}$ to $10^{-4}$ $M$, at least three species have an $\alpha$ of at least 0.1, although it is a different three at each value of log [$NH_3$]. Note also that the line for any species only becomes straight at very low $\alpha$ values.

Figure 3-4 illustrates an alternative method of graphing the same information. In this graph the $\alpha$ values are stacked up, so to speak, so that their total (which must always be 1.00) fills a given vertical distance. For example, at log [$NH_3$] $= -3$, $\alpha_0 = 0.01$, $\alpha_1 = 0.14$, $\alpha_2 = 0.46$, $\alpha_3 = 0.35$, $\alpha_4 = 0.04$, and $\alpha_5$ is a very small number. $\alpha_0$ is graphed as the distance above the baseline; $\alpha_1$ is graphed as a distance above $\alpha_0$, $\alpha_2$ as a distance

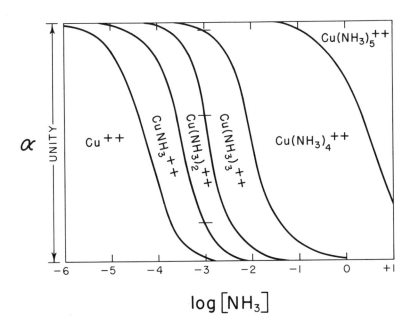

**Fig. 3-4.** An alternative form of distribution graph (compare with Fig. 3-3). The vertical distance between any two lines is proportional to the fraction of total Cu(II) present as the indicated species at any chosen log [$NH_3$]. See text.

above $\alpha_1$, etc., until the entire distance from the baseline to the top of the graph is filled in. For the example of $10^{-3}$ $M$ NH$_3$, these distances are indicated by the small horizontal lines above $-3$ on the log [NH$_3$] scale. When the points for each are connected over a range of log [NH$_3$], the smooth curves shown in Fig. 3-4 are obtained. The value of $\alpha$ for any species is graphed then as a vertical distance between two lines, and is so labelled in Fig. 3-4. Since the lines themselves are nearly vertical between $\alpha = 0.1$ and $0.9$, the graph is also divided horizontally into regions of log [NH$_3$] within which the indicated species predominates.

Note the relatively wide range in Figs. 3-3 and 3-4 within which Cu(NH$_3$)$_4^{++}$ predominates. This is, of course, a consequence of the gap between $Q_4$ and $Q_5$, which in turn reflects a preferred symmetry for bonding in Cu$^{++}$. The Cu$^{++}$ ion has 9 $d$ electrons, one short of a filled $d$ subshell. This may be thought of as a positive hole in an otherwise spherical cloud of negative charge. If this positive "hole" can be placed in the $d$ orbital, which points directly at the electron-rich NH$_3$ ligands, repulsion between the ligands and the $d$ subshell electrons can be minimized. If the first four ligands are thought of as lying on the $x$ and $y$ axes, with Cu$^{++}$ at the origin, the positive hole occupies the $d_{x^2-y^2}$ orbital (Fig. 3-5). The electron pair of a fifth ligand must face a fully occupied set of orbitals on or near the $z$ axis (the $d_{xz}$, $d_{yz}$, and $d_{z^2}$); hence the large numerical gap between $Q_4$ and $Q_5$.

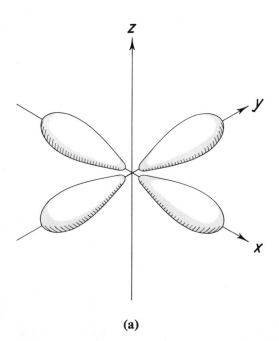

**(a)**

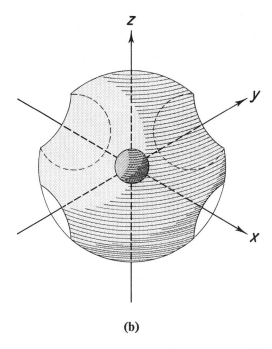

**(b)**

**Fig. 3-5.** (a) A $d_{x^2-y^2}$ orbital. (b) A metal ion (dark sphere) surrounded by a spherical shell containing a hole the shape of a $d_{x^2-y^2}$ orbital. This hole is empty in the case of $Ni^{++}$ and half-filled in the case of $Cu^{++}$.

EXERCISE 3.4:   The reaction between $Cu^{++}$ and $NH_3$ is sometimes written in elementary texts as

$$Cu^{++} + 4NH_3 = Cu(NH_3)_4^{++}$$

On the basis of Fig. 3-4, how justified is this equation? What is the predominant species in 1 $M$ $NH_3$? In 6 $M$? What would result if $Cu^{++}$ and $NH_3$ were mixed in exact 1-to-4 ratio?

## Slightly Dissociated Salts

Complex formation between the cation and the anion of some salts causes them to be incompletely dissociated in solution. The most famous examples are the halides of $Hg^{++}$, but many other salts often assumed to be completely dissociated in water solution form anion-cation complexes with measurable stability constants. Most such salts contain ions of high charge; the transition and alkaline earth metals, and sulfate and phosphate salts,

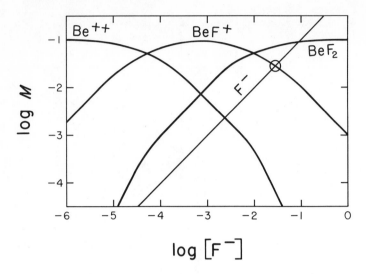

**Fig. 3-6.** A logarithmic concentration distribution graph for the $Be^{++}$-$F^-$ complexes in aqueous solution. The circled intersection represents the situation in the absence of added $F^-$.

present many examples. Figure 3-6 is a distribution graph for the salt, beryllium fluoride, giving the concentrations of $Be^{++}$ and its fluoro complexes, and of $F^-$ in a 0.1 $M$ solution of $Be^{+2}$ ($[F^-]$ is assumed arbitrarily variable, as $[H^+]$ is in a pH distribution graph). The stability constants of the complexes are $K_1 = 10^{4.3}$, $K_2 = 10^2$. Note first that, in the presence of excess $F^-$, the salt is practically undissociated; in 1 $M$ $F^-$, only 1% is present as $BeF^+$, and 99% is $BeF_2$.

We may also read from Fig. 3-6 the state of affairs in a 0.1 $F$ $BeF_2$ solution, in the absence of added $F^-$. A mass balance on $F^-$ (analogous to a proton balance in a Brønsted equilibrium) gives

$$[F^-] = [BeF^+] + 2[Be^{++}] \tag{3-5}$$

The highest crossing in Fig. 3-6 of $[F^-]$ with any term on the $F^-$-deficient side of Eq. 3-5 is the $F^-$-$BeF^+$ crossing at log $[F^-] = -1.6$; at this point (circled in Fig. 3-6), $[Be^{++}]$ is less than $10^{-4}$ $M$. This is negligibly small compared to $[BeF^+]$, which is, of course, also $10^{-1.6}$ $M$. Converting to ordinary numbers, we find that $[BeF^+] = [F^-] = 3 \times 10^{-2}$ $M$, and $[BeF_2] = 7 \times 10^{-2}$ $M$. The salt is only 30% ionized in the first step, and less than 1% ionized in the second step.

We should expect that incomplete dissociation would be prevalent among "2-2" salts, and this is indeed the case. $Zn^{++}$ and $SO_4^{--}$ form a complex with a stability constant of $10^{2.31}$, or 203, and 1 $F$ $ZnSO_4$ is thus only about 7% dissociated.

EXERCISE 3.5:  Assuming no species present other than $Zn^{++}$, $SO_4^{--}$, and $ZnSO_4$ (solute), derive an equation for the fraction of the total Zn present as the uncomplexed ion $Zn^{++}$ and verify this calculation. How does the fraction dissociated vary with the formal concentration of $ZnSO_4$? Where have we seen this sort of thing before?

The extent to which even 1-1 salts like NaCl are dissociated in concentrated solutions is still a subject for controversy and investigation. The continuum from covalent to ionic bonding in complexes may be extended to weaker and weaker coulombic interactions until it is difficult to decide whether to call the interaction between very weak, hard acids and bases a chemical bond or a "merely" coulombic one (a fundamentally meaningless distinction for any interaction between atoms anyway). It is clear that in concentrated solutions ions of opposite charge may share the same shell of oriented solvent molecules, thus forming an "ion pair." However, any distinction between this situation and the lowering of the "separated" ions' activities through "ordinary" ionic-strength effects[2] becomes more arbitrary and experimentally unrealizable the more concentrated the solution.

## Metal-Chelon Equilibria

Though polydentate ligands have been known and used in chemistry for many decades, a real renaissance of interest in the chemistry of metal complexes came with the invention and synthesis of aminopolycarboxylic acids of which the archetype is ethylenediaminetetraacetic acid (EDTA). These ligands have sufficient basic sites on one molecule to accommodate the maximum coordination number of most metals, and, if cleverly designed, are of a proper size and geometry to wrap around a metal ion, virtually engulfing it in Lewis basicity and forming bonds with the octahedral symmetry that many metal ions allow (Fig. 3-7). The result is that the stoichiometry of a reaction between a metal ion and a large and complex chelon is likely to be one-to-one, which greatly simplifies equilibrium calculations and, as we will see, has an important effect on its equilibrium constant. The exceptions to this rule are generally bi- or tridentate ligands like the oxalate ion or triaminocyclohexane.

The fact that metal chelates have large stability constants, which is an important advantage in many applications, is somewhat surprising at first glance. For example $Mg^{++}$, which forms weak complexes with $NH_3$ (log $K_1 = 0.23$, log $K_2 = -0.15$) and acetate ions (log $K_1 = 0.82$), forms an EDTA chelate with log $K = 8.69$. Clearly some factor present when all the basic sites are on one molecule does not operate in the reaction of these

[2]See Appendix 2.

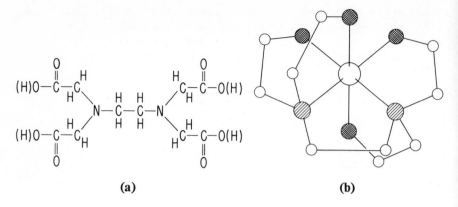

**(a)**                      **(b)**

**Fig. 3-7. (a)** Ethylenediaminetetraacetic acid (EDTA). For bonding to a metal ion, the parenthesized protons must be ionized, leaving the EDTA$^{4-}$ ion. The chelon may then be a hexadentate ligand, bonding through the two nitrogens and four carboxyl oxygens. Not all metal-EDTA complexes utilize all six basic sites on the chelon, however. **(b)** The geometry of a metal ethylenediaminetetraacetate which involves all of the basic sites on the chelon. ◯ , metal ion; ◨, N; ◍, O; ◯, C. The hydrogens on the carbon chains and one of the oxygens on each carboxyl group are omitted for clarity.

same basic groups as individuals. Part of the answer is that in reacting with EDTA, Mg$^{++}$ is able to form up to six bonds at once, with a consequent lowering of energy that is several times as advantageous as that obtained in forming only one bond. However, this is not the only factor. The stability of the complex Mg(NH$_3$)$_2$(acetate)$_4^{--}$ which would correspond to disjointed MgEDTA$^{--}$ has not been measured (and is probably immeasurably small), but consider the series:

$$Cd^{++} + 4H_2NCH_3 = Cd(H_2NCH_3)_4 \qquad Q_1Q_2Q_3Q_4 = 10^{6.55}$$

$$Cd^{++} + 2H_2NCH_2CH_2NH_2 = Cd(H_2NCH_2CH_2NH_2)_2 \qquad Q_1Q_2 = 10^{10.02}$$

$$Cd^{++} + N(CH_2CH_2NH_2)_3 = CdNCH_2CH_2NH_2 \qquad Q = 10^{12.3}$$

In each of these overall reactions, four methylated nitrogens bond to the cadmium ion. The numerical values of the overall stability quotients cannot be compared directly, because they have different units. For example, the units of $Q_1Q_2Q_3Q_4$ are liters$^{+4}$ moles$^{-4}$, whereas the units of $Q$ are liters/mole (you should convince yourself of this). Just as you cannot say how many miles there are in an acre, you cannot convert liters/mole to liters $^{+4}$ moles $^{-4}$. However, we can calculate the quantity of uncomplexed

$Cd^{++}$ in equilibrium with 1 $F$ excess ligand in each case, and find that the ratio of $Cd^{++}$ to fully complexed Cd is $10^{-6.5}$, $10^{-10.02}$, and $10^{-12.3}$, respectively, in the three cases. Thus in a loose way, we may say that the complexes are more stable, the smaller the number of particles out of which they are made. The situation resembles order and chaos (the more pieces a complex can break into, the smaller the fraction of unbroken complex present).

It is not quite fair to let matters rest with this simple argument, because we should remember that water molecules bonded to the $Cd^{++}$ must depart when the nitrogen bases arrive, and the order-chaos balance should include this effect. There is also good evidence that, at least in the case of chelons containing a number of double bonds (such as aromatic molecules and carbonyls), metal $d$ electrons may be considerably delocalized into the ring formed (Fig. 3-8) when two sites on the same molecule bond to the same metal ion. This may lead to a qualitatively different and stronger sort of bonding than is possible when only one bond is formed to each ligand.

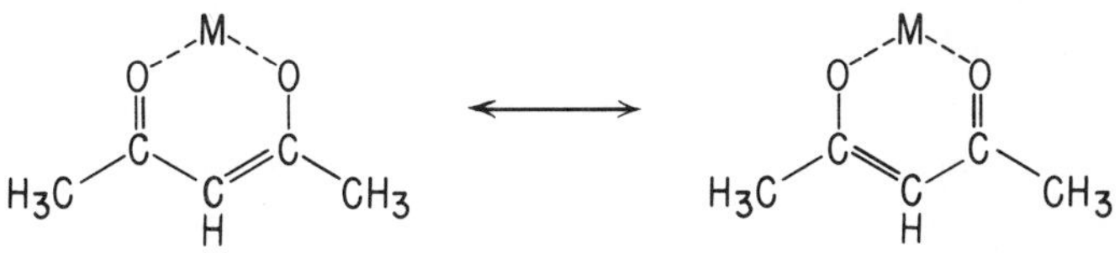

**Fig. 3-8.** "Chelate rings" formed when a bidentate ligand (in this case, the acetylacetonate negative ion) bonds to a metal. The indicated delocalization of bonding electrons may include the metal in some cases, and may lead to unexpectedly large stability constants for the complex (see text).

The fact that chelons form stable complexes with many metal ions has made them attractive to analytical chemists, who are always on the lookout for reactions with large equilibrium constants, since these are usually necessary for successful analytical procedures. For example, before the advent of EDTA, a rapid, convenient analysis for $Mg^{++}$ and $Ca^{++}$, both of which are ubiquitous in nature, was difficult to come by. Now they may be titrated, together or separately, using as an indicator a chelon such as eriochrome black T (Fig. 3-9), which changes color when it complexes with a metal ion. It has been analytical chemists, by and large, who have elucidated the complex equilibria which lie behind the attractively simple 1-to-1 stoichiometry of most metal chelates.

**Complications in metal-chelon equilibria.**    It is all very well to induce a normally unreactive $Ca^{++}$ ion to form a complex by loading a lot of Lewis basic sites onto one molecule; but we should remember that all Lewis bases

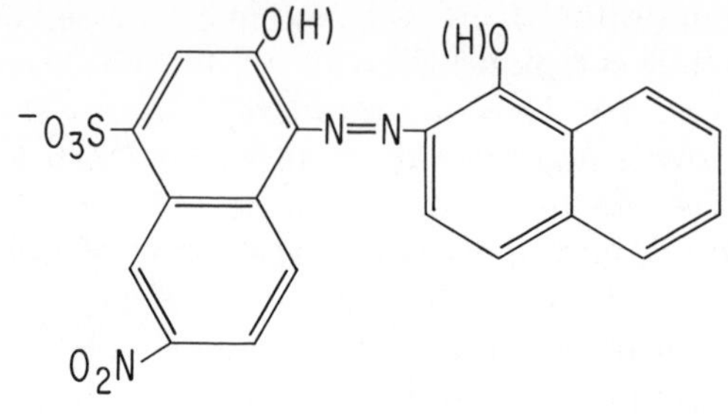

**Fig. 3-9.** The structure of eriochrome black T. The parenthesized protons are lost before a metal-EBT chelate is formed.

are also Brønsted bases, since the proton is just one example of a Lewis
acid. As a result, the deprotonated form of the ligand which reacts with
metal ions, for example $EDTA^{4-}$, is a Brønsted base of very respectable
basic strength. To avoid the protonation of the Lewis basic sites on the
EDTA anion, we should expect to have to keep the pH of the solution
quite high. Because water is a protolytic solvent, there is an unavoidable
penalty for minimizing protons; we simply maximize $OH^-$ ions and these,
in addition to being Brønsted bases, are good Lewis bases, particularly
for metal ions. They are, therefore, competition for the chelon and are also
to be avoided. If we buffer the solution to strike a compromise between too
many protons and too many hydroxide ions, we run into a third hazard:
since we can only buffer a solution at a moderate pH through the presence
of a weak Brønsted acid and its conjugate *base*, we have to have present
another electron-rich species. For the metal ion $M^{+n}$ being reacted with
$EDTA^{4-}$ in a solution buffered by the conjugate pair $HX - X^-$, we have
the following general equilibrium:

$$M(OH)_n(c)$$

$$\Updownarrow$$

$$nOH^-$$
$$+$$

$$MOH^{n-1} \rightleftharpoons OH^- + \boxed{M^{+n} + EDTA^{4-} \rightleftharpoons MEDTA^{n-4}} \qquad (3\text{-}9)$$

$$\Updownarrow \qquad\qquad\qquad + \qquad\quad +$$

$$M(OH)_2^{n-2} \qquad\qquad\qquad X \qquad\quad H^+$$

$$. \qquad\qquad\qquad\qquad \Updownarrow \quad\; \Updownarrow$$

$$. \qquad\qquad\qquad\qquad MX \quad\; HEDTA^{3-}$$

$$. \qquad\qquad\qquad\qquad + \qquad\quad +$$

$$X \qquad\quad H^+$$

$$\Updownarrow \quad\; \Updownarrow$$

$$MX_2 \quad\; H_2EDTA^{-2}$$

$$. \qquad\quad .$$

$$. \qquad\quad .$$

$$. \qquad\quad .$$

If the equilibrium within the box is the only one we are interested in promoting, we must accept the presence of the others as necessary accompaniments. As we will see, they need not be evils, because they offer a means of controlling the otherwise voracious reactivity of EDTA toward metal ions.

The approach taken by analytical chemists in unraveling Eq. 3-9 is to define a *conditional stability quotient* $Q'$, which takes all of the interfering equilibria into account:

$$Q' = Q_{MEDTA}\alpha_{EDTA}\alpha_M \qquad (3\text{-}10)$$

where

$$\alpha_{EDTA} = \frac{[EDTA^{4-}]}{[EDTA^{4-}] + [HEDTA^{3-}] + [H_2EDTA^{2-}] + \cdots} \qquad (3\text{-}11)$$

and

$$\alpha_M = \frac{[M]}{[M] + [MOH] + [M(OH)_2] + \cdots + [MX] + [MX_2] + \cdots} \qquad (3\text{-}12)$$

(Since the charges on M and X are arbitrary, we will drop them in what follows.) That is, if we define $C_{EDTA}$ as the formal concentration of the chelon not reacted with M and $C_M$ as the formal concentration of M not reacted with the chelon, $\alpha_{EDTA}$ is the fraction of $C_{EDTA}$ present as unprotonated $EDTA^{4-}$, and $\alpha_M$ is the fraction of $C_M$ present as the "free" hydrated ion $M^{+n}$.

EXERCISE 3.6:   Show that, from the definitions of $Q'$, $\alpha_{EDTA}$, $\alpha_M$, $C_{EDTA}$, and $C_M$, it follows that $Q' = [MEDTA]/C_M C_{EDTA}$. (This relationship gives the conditional stability quotient its name.)

The use of the conditional stability quotient allows one to predict the net reactivity of the metal ion with the chelon in the presence of the competing side reactions of 3-9. An important use of $Q'$ is to devise conditions under which only one metal ion will react with the chelon in the presence of many others.

For example, most transition metal ions form stable cyanide complexes, whereas the alkaline earth metals do not. By adding NaCN to a mixture of, say, $Ca^{++}$ and $Cd^{++}$, $\alpha_{Cd}$ is reduced to a very small number (roughly $10^{-19}$ in 1 $M$ $CN^-$), and $\alpha_{Ca}$ is unaffected. Then $Q'$ for $CdEDTA^{--}$ is decreased by a factor of $10^{-19}$ and $Q'$ for $CaEDTA^{--}$ remains large, allowing $Ca^{++}$ to be titrated with EDTA in the presence of $Cd^{++}$, even though the "absolute" stability quotient of $CdEDTA^{--}$ is larger than that of $CaEDTA^{--}$ by a factor of $10^6$. The $Cd^{++}$ is sometimes said to be "masked" by the presence of the competing ligand $CN^-$; as far as reaction with EDTA goes, it might as well not be there, because it is hidden away from the action in a cloud of soft Lewis basicity.

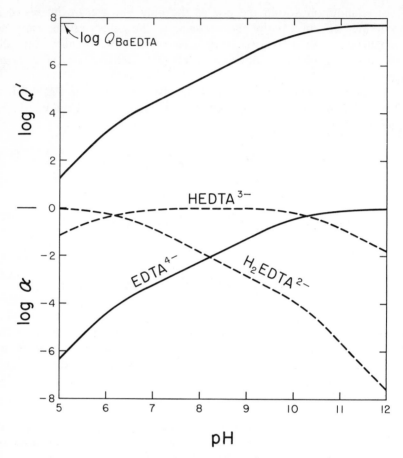

**Fig. 3-10.** Lower curves: a logarithmic distribution graph for the three most basic forms of EDTA. Upper curve: the conditional stability quotient $Q'$ for the BaEDTA$^{--}$ complex as a function of pH. Note the parallelism of $Q'$ and $\alpha_{EDTA}$, which is a consequence of the definition of $Q'$ (see text).

The calculation of $\alpha_{EDTA}$ and $\alpha_M$ presents nothing new to us at this point. We have encountered these calculations in Chapter 2 and earlier in this chapter. The only information required is the pH and the stepwise dissociation constants of the fully protonated ligand (for $\alpha_{EDTA}$), the concentration of competing Lewis bases such as OH$^-$ and X, and the relevant M-OH$^-$ and M-X stability constants. The effect of protonation alone on $Q'$ of BaEDTA$^{--}$ is shown in Fig. 3-10. Note that log $Q'$ remains parallel to log $\alpha$ of EDTA$^{4-}$, and approaches log $Q$ when log $\alpha$ of EDTA$^{4-}$ approaches zero. Some further practice is available in the problems at the end of this chapter.

## Binuclear Complexes

The hexaquo $Fe^{3+}$ ion has a considerable degree of paramagnetism, since it has five unpaired electrons in the $3d$ subshell. A rather puzzling aspect of this phenomenon is that the magnetism of the ion appears to be pH-dependent, decreasing as the pH is increased from the very low pH region in which $Fe(H_2O)_6^{3+}$ exists. A plausible guess is that the formation of hydroxy-complexes is taking place, but this reaction is insufficient to account for the loss of magnetism. However, careful equilibrium studies show that the complex species formed depends not only on the ligand concentration as it should for an ordinary complex series such as $Cu^{++}$—$NH_3$, but on the formal concentration of $Fe^{3+}$ as well. The latter is good evidence for the existence of a *polynuclear* complex (that is, one which contains more than one metal atom).

EXERCISE 3.7:   How does the last sentence follow? Where have we encountered a similar situation?

The species that fits the equilibrium data has the formula $Fe_2(OH)_2^{4+}$; that is, it is a dimer of $FeOH^{2+}$. The proposed structure is

$$
\begin{array}{c}
H \\
| \\
O \\
\diagup \quad \diagdown \\
(H_2O)_4Fe \qquad Fe(H_2O)_4 \\
\diagdown \quad \diagup \\
O \\
| \\
H
\end{array}
$$

the entire entity having a charge of $+4$. It has been suggested that interaction of the $d$ electron spins through the hydroxy bridges accounts for the loss of paramagnetism observed when this species begins to predominate in the solution. *Bridging* ligands like the OH groups in this complex are being found with increasing frequency. They generally entail the use of what we ordinarily think of as nonbonding electrons on electronegative atoms. Another example is the ion,

$$
\begin{array}{c}
O - O \qquad ^{4+} \\
\diagup \quad \diagdown \\
(en)_2Co \qquad Co(en)_2 \\
\diagdown \quad \diagup \\
N \\
| \\
H_2
\end{array}
$$

where en = ethylenediamine, $H_2N(CH_2)_2NH_2$.

Because the composition of a solution containing binuclear complexes depends on concentrations of both the ligand and the metal ion, a mathematical treatment of these systems is rather complex, and we will leave it to more advanced treatments, some of which are listed in the bibliography at the end of this chapter.

## Lewis Acid-Base Reactions in Medicine

Since many physiologically important molecules depend for their function on electron-rich atoms or groups, it is not surprising that Lewis acid-base interactions play an important role in the chemistry of life, and that accidental ingestion of Lewis bases or acids outside the bounds of normal body tolerances can be dangerous indeed. Interestingly, hard acids and bases tend to be slightly less dangerous than soft ones. Arsenic (in the +3 state a soft acid) and cyanide (a soft base) are staples on the shelves of mystery writers, whereas the otherwise rather similar $Al^{3+}$ (probably ingested more than is realized from aluminum cooking utensils) and chloride ions, which are, respectively, a hard acid and a hard base, are relatively harmless. Carbon monoxide, a soft base which forms very stable complexes with metal atoms in low or zero oxidation states, voraciously attacks the iron of hemoglobin by bonding at its less electronegative carbon end, but

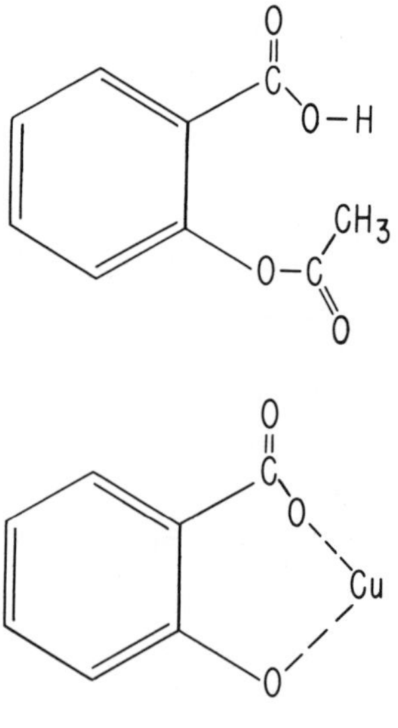

Fig. 3-11. The structure of aspirin, as found in the pill (upper structure) and as altered in the digestive tract to the lower structure, whose possibility as a ligand for $Cu^{++}$ is illustrated.

$N_2$, which is isoelectronic and identical to CO except for the shift of a positive charge from the oxygen nucleus to the carbon nucleus, is hard on both ends and harmless to life. $Ag^+$ is soft and poisonous, $Na^+$, about the same size, is hard and harmless. In fact, most of the metal ions which are poisonous in small quantities — $Ag^+$, $Hg^{++}$ and $Hg_2^{++}$, $Pb^{++}$, $Bi^{3+}$, and their neighbors — are rich in $d$ electrons and are soft acids.

Recently the treatment of heavy metal poisoning has been revolutionized by the use of chelons as medicines (or reagents, if you like) to complex the metal ions away from whatever body chemistry they are involved in, get them into the bloodstream as soluble species, and out of the body through the kidneys. EDTA is especially useful in lead poisoning (increasingly prevalent in our increasingly polluted world), although it is not effective against mercury, which is amenable to treatment with other chelons whose mercury complexes have larger stability constants. Many of the family of wonder drugs such as aureomycin and the like are good chelons, and it appears that aspirin works its wonders through its specific ability to control the level of $Cu^{++}$ in the bloodstream. The structure of aspirin, and its possibilities as a chelon, are shown in Fig. 3-11.

Not all chelons are beneficial or even harmless, however; for example, 3-pentadecylcatechol

causes blistering and itching on contact with the skin, and has been known to incapacitate victims who unwisely handled the plant (*rhus toxicodendron*) whose sap contains it. The best therapy against this toxic chelon is an ointment containing a good Lewis acid; salts of Zr(IV) are frequently used. An even better practice is to avoid contact with *rhus toxicodendron*, which is easily recognized by its three leaves growing from a woody stem.

## Suggestions for Further Reading

Butler, James N., *Ionic Equilibrium: A Mathematical Approach*. Reading, Mass.: Addison-Wesley Publishing Co., 1964. Chapters 8, 9, and 10 cover Lewis acid-base equilibria, including polynuclear complexes.

Kolthoff, I. M., P. J. Elving, and E. B. Sandell, eds., *Treatise on Analytical Chemistry*. New York: Interscience Encyclopedia, 1959. Part I, Vol. 1, Ch. 14.

Pearson, R. G., "Soft and Hard Acids and Bases." *Science, 151*, 1721 ff. (1966).

Schubert, Jack, "Chelation In Medicine." *Scientific American*, May 1966, pp. 40–50.

## Problems

*(Use data from Appendix 3 where necessary).*

3.1. Identify the Lewis acid and Lewis base from which each of the following species might have been formed:
   A. $(CH_3)_3NBF_3$
   B. $CuCl^+$
   C. $NH_4^+$
   D. $H_2O$
   E. $AgS_2O_3^-$
   F. $Co(en)_2ClH_2O^{++}$

3.2. Draw distribution graphs of the form of Figs. 3-3 and 3-4 for the $Hg^{++}$-$Cl^-$ series of complexes.

3.3. From the logarithmic distribution graph for Problem 3.2 and a chloride balance, show that the predominant form of mercuric chloride dissolved in pure water is the molecule $HgCl_2$. What fraction of the total mercury is present in all other forms combined when $C_{HgCl_2} = 1$ $F$?

3.4. A. Construct a logarithmic distribution graph for the complexes of $Fe^{3+}$ with $SCN^-$
   B. The formation of one or more of the colored $Fe^{3+}$-$SCN^-$ complexes is used as a qualitative test for the presence of Fe(III) in a solution. What species predominate when 1 ml of 0.1 $F$ KSCN is added to 10 ml of a water sample containing $10^{-5}$ $M$ Fe(III)?

3.5. The chelon NTA (nitrilotriacetic acid, $N(CH_2COOH)_3$) is a triprotic acid with dissociation quotients $pQ_1 = 1.9$, $pQ_2 = 2.5$, and $pQ_3 = 9.7$.
   A. In what pH range is it most effective as a chelon? Sketch $\alpha_{NTA}$ as a function of pH.
   B. The stability quotient of $SrNTA^-$ is $10^{5.0}$; that of $ZnNTA^-$ is $10^{10.5}$. Using data from Appendix 3, select a masking agent and describe conditions under which $Zn^{++}$ may be masked sufficiently so that the *conditional* stability quotient of $SrNTA^-$ is larger than the conditional stability quotient of $ZnNTA^-$ by a factor of $10^5$.

3.6. The stability quotient of the $Mg^{++}$-eriochrome black T (EBT, Fig. 3-9) complex is $10^{7.0}$.
   A. What is its conditional stability quotient at a pH of 10.0 if $pQ_{b(1)}$ of $EBT^{3-}$ is 0.5, and $pQ_{b(2)}$ is 6.6?
   B. At a pH of 10.0, what is the value of $C_{Mg}$ such that $[MgEBT^-] = C_{EBT}$? (This is the approximate condition for a color change in a titration of $Mg^{++}$ using EBT as an indicator.)

# 4 Solubility Equilibria

## Introduction

Having devoted three chapters to a single phase, let us now add another ingredient: a second pure phase in equilibrium with a solution. For example, suppose we place a chunk of solid $I_2$ at the bottom of a container of pure water. We will see yellowish color sluggishly spreading out from the $I_2$ and, after a day or so, a uniform, moderately dark yellow solution of $I_2$ in water. Most of the solid $I_2$ will remain unchanged. The equilibrium we have established is

$$I_2(c) = I_2(s) \qquad (4\text{-}1)$$

where (c) refers to the crystalline state of solid iodine, and (s) refers to iodine as solute.

The equilibrium quotient for Eq. 4-1 is

$$Q_s = [I_2(s)] \qquad (4\text{-}2)$$

Since $I_2(c)$ is a pure phase, we use mole fraction as its concentration unit and need not include its value of unity in the expression for $Q_s$. Indeed, because there is no need to identify which concentration we are referring to in the $Q_s$, we will usually omit the (s).

Equation 4-2 embodies a familiar idea: when solubility equilibrium is reached, a constant $[I_2]$ characterizes the solution which, since it holds this much and neither more nor less, is said to be *saturated* with $I_2$. Except for changes in the value of $Q_s$ with temperature, Eq. 4-2 says all there is to be said (at least algebraically) about the solubility of iodine in pure water. Note that nowhere in the discussion of Brønsted or Lewis acid-base equilibria have we encountered an equilibrium quotient quite this simple. The simplicity arises from the fact that the presence of a second phase in equilibrium with the solution puts an additional restraint on the system which, in this case, is sufficient to fix its composition exactly. In water at 25°C, the value of $Q_s$ for $I_2$ is $1.32 \times 10^{-3}$ $M$.

While things are still relatively simple, we will make a seemingly obvious definition that will be useful to us in more complicated situations. The *solubility* of a substance we will define to be the number of formula weights of the pure phase which must dissolve in order to form one liter of a saturated solution. In the case of $I_2$ dissolving in pure water, this is clearly equal to $[I_2]$ and to $Q$.

However, things can easily become less simple. Iodine can interact with iodide ions in a Lewis acid-base reaction to form the species $I_3^-$, the triiodide ion:

$$I_2 + I^- = I_3^- \qquad (4\text{-}3)$$

If we allow solid $I_2$ to come to equilibrium with a solution containing $I^-$, then $I_2$ molecules entering the solution may meet one of two fates: They will either exist (at any particular instant) as $I_2$ or as part of an $I_3^-$ ion. By our definition of solubility, we count both:

$$S = [I_2] + [I_3^-] \qquad (4\text{-}4)$$

If we formulate the equilibrium this way:

$$\begin{array}{c} I_2(c) = I_2(s) \\ + \\ I^- \;\; = \;\; I_3^- \end{array} \qquad (4\text{-}5)$$

we see that $I_2$ in solution is still in equilibrium with the solid, so that Eq. 4-2 still applies. The concentration of molecular iodine $I_2$ is still equal to $Q$, but the solubility is now larger than either.

EXERCISE 4.1:   If the equilibrium $[I^-]$ in a system like (4-5) is known, there are three unknowns ($S$, $[I_2]$, and $[I_3^-]$) in a saturated solution. From mass balance (Eq. 4-4), the solubility quotient $Q$, and the formation quotient $Q_f$ of the $I_3^-$ complex, derive an equation for the solubility as a function of the $I^-$ concentration. Sketch a graph of $S$ as a function of $[I^-]$.

**Solubility rules.**   In our consideration of ionic solutions in Chapter 1 we established a simple guiding principle for understanding the approximate values of solubility equilibrium constants. All things being equal, a solution will form unless a large input of energy is required to overcome inter-molecular forces in either the solute or the solvent. For example, there is no upper limit (except unit mole fraction) to the solubility of room tempera-ture benzene in room temperature cyclohexane, because the forces between molecules in the solvent and the solute (if such a distinction is meaningful in this case) are so nearly equal that disorder carries the day.

If the prospective solute has large internal forces between particles, as in the case of ionic solids, then solutions form only to the extent that these forces are weakened by a high dielectric constant in the solvent and compensated for by strong solvent-solute interactions. Finally, if the solute has only weak interparticle forces, but the solvent has strong ones which are not compensated for by solvent-solute interactions, the solubility is again low; an example is the $I_2$-$H_2O$ system we have been considering.

All of these rules can be summarized in a single one: *Like dissolves like,* where by "like," we mean similar in polarity. Iodine dissolves much more readily in ethanol ($Q_s = 0.97$) than in water, which accounts for the familiar *dark* brown ethanol solution of $I_2$ used to sterilize cuts. On the other hand sugar, with all its polar, hydrogen-bondable OH groups, dis-solves much more readily in water than in gasoline.

## Ionic Solubility

When the solute dissociates into ions, there is more room for the system to maneuver, as we can see from this typical equilibrium:

$$PbSO_4(c) = Pb^{++} + SO_4^{--} \qquad (4\text{-}6)$$

The equilibrium quotient for such a reaction is called a *solubility product,* because the concentration unit for the pure phase is mole fraction, and its value of unity does not appear:

$$Q_{sp} = [Pb^{++}][SO_4^{--}] \qquad (4\text{-}7)$$

The requirement of the equilibrium quotient is looser here, because there is an enormous number of combinations of $[Pb^{++}]$ and $[SO_4^{--}]$ whose product is equal to the $Q_{sp}$. Of course, if the system consists only of lead sulfate and water, $[Pb^{++}]$ must equal $[SO_4^{--}]$. In that case, from Eq. 4-7,

$$S = [Pb^{++}] = [SO_4^{--}] = (Q_{sp})^{1/2} \qquad (4\text{-}8)$$

Note well that Eq. 4-8 is valid *only* if there is no excess of either $Pb^{++}$ or $SO_4^{--}$ over that supplied by the solubility of solid $PbSO_4$. Such a condition only arises when pure $PbSO_4$ is added to a solution containing no $Pb^{++}$ or $SO_4^{--}$, or when a $Pb^{++}$ salt and a $SO_4^{--}$ salt are mixed in equal quantities. Equation 4-8 is such a simple and neat result that the temptation to write it down under any and all circumstances appears to be nearly irresistible. Suppress it.

**Stoichiometries other than 1-to-1.**    If the cation and anion of the salt do not have equal and opposite charges, then the $Q_{sp}$ of course must reflect the unequal stoichiometry of the salt, and the solubility in pure water (analogous to Eq. 4-8) takes on a slightly more complex form. For the salt $M_aX_b$, the solubility equilibrium

$$M_aX_b(c) = aM + bX \qquad (4\text{-}9)$$

gives the $Q_{sp}$ the form

$$Q_{sp} = [M]^a[X]^b \qquad (4\text{-}10)$$

(The ionic charges are omitted for clarity.) In a solution containing no excess of either M or X, the solubility in terms of formula weights of $M_aX_b$ (that is, in terms of our definition) is

$$S = \frac{[M]}{a} = \frac{[X]}{b} \qquad (4\text{-}11)$$

The $Q_{sp}$ may be written as

$$Q_{sp} = (aS)^a(bS)^b$$
$$= a^ab^bS^{(a+b)} \qquad (4\text{-}12)$$

so that

$$S = \left(\frac{Q_{sp}}{a^ab^b}\right)^{1/(a+b)} \qquad (4\text{-}13)$$

EXERCISE 4.2:   Apply to Eq. 4-13 the numbers for $Ag_2CrO_4$, $CaF_2$, and $Ga_2Te_3$. Does the case of $PbSO_4$ already treated specifically also fit Eq. 4-13?

Actually, real cases of complex stoichiometry in which Eq. 4-13 accurately represents the solubility in pure water are rarer than books such as this pretend. For complex stoichiometry, the ions involved must have high charge, and are thus likely to be good Lewis acids or bases; as a result, Eq. 4-9 does not describe all of the equilibria involved. We will investigate the effect of secondary equilibria on ionic solubility in due course.

**The common-ion effect.**   Among numerous situations in which Eqs. 4-8 or 4-13 are not valid is the presence of either M or X ions in solution from some source other than $M_aX_b$. For example, the sulfate concentration of a solution is sometimes determined by adding excess $BaCl_2$ to it to form $BaSO_4(c)$, which is filtered out, dried, and weighed. Suppose that to a solution initially 0.01 $M$ in $SO_4^{--}$, a threefold molar excess of $BaCl_2$ is added so that there is present in solution 0.02 $M$ *excess* $Ba^{++}$. Equilibrium 4-9 must clearly shift to the left, and $BaSO_4$ precipitates out. We can no longer equally well measure the solubility of $BaSO_4$ in the solution by $[Ba^{++}]$ or by $[SO_4^{--}]$, since the $Ba^{++}$ present comes from the added $BaCl_2$, as well as from the $BaSO_4$. However, the equilibrium mixture of $BaSO_4(c)$, $Ba^{++}$, and $SO_4^{--}$ is the same as that obtained by equilibrating pure solid $BaSO_4$ with 0.02 $F$ $BaCl_2$. For every mole of $BaSO_4$ that dissolves, one mole of $SO_4^{--}$ appears in solution, and one mole of $Ba^{++}$ is added to the original quantity of $Ba^{++}$ there. Algebraically,

$$[SO_4^{--}] = S \qquad\qquad (4\text{-}14)$$

and

$$[Ba^{++}] = S + 0.02 \qquad\qquad (4\text{-}15)$$

In the general symbology of Eqs. 4-11 and 4-12,

$$[X] = bS$$

and

$$[M] = aS + C$$

where $C$ is the molarity of the *excess* salt present. Then for a general problem, we may say

$$Q_{sp} = (aS + C)^a(bS)^b \qquad\qquad (4\text{-}16)$$

which we could expand into an exact but unpalatable equation of degree $a + b$.

However, in preceding chapters, we have generally found that practical problems allow for simpler approximate solutions when chemically reasonable approximations are made. In the present case, we are investigating the solubility of a slightly soluble substance; and that solubility, furthermore, is being repressed by the presence of a common ion. That is, $S$ is quite likely to be a small number, and high powers in $S$ will be much smaller than $S$ itself. As an example, let us return to the $BaSO_4$ problem. For the conditions of this problem (Appendix 2), $Q_{sp} = 6 \times 10^{-10}$. Then Eq. 4-16 becomes

$$6 \times 10^{-10} = (S + 0.02)(S)$$
$$= S^2 + 0.02S$$

Momentarily discarding the $S^2$ in this equation, we find

$$6 \times 10^{-10} = 0.02S$$
$$S = 3 \times 10^{-8} M$$

Given this result, we can see that $S^2$ is certainly quite small compared to $0.02S$, and our assumption was a good one.

> EXERCISE 4.3:    What change in the mass balance (such as Eq. 4-15) would be required to eliminate all powers of $S$ higher than $b$ in Eq. 4-16? Give a verbal explanation of this change and a semiquantitative justification. What if, in the $BaSO_4$ problem, $C$ had been $1 \times 10^{-4} F$?

Since by ignoring higher powers in $S$ we are invariably overestimating $S$ (by a tiny fraction in most cases), we may be confident that such approximate solutions will be valid as long as $S$ is quite small compared to $C$ (since then, for example in the $BaSO_4$ problem, $S^2$ will be small compared to $CS$). When the approximate solution for $S$ is no longer small compared to $C$, we must include higher powers in $S$; as long as $S$ is less than $1$ $M$, however, it is still true that the importance of a term in $S$ will drop off sharply with increasing powers — an important consideration if it allows the substitution of a relatively simple quadratic for a cubic or higher degree equation in $S$.

> EXERCISE 4.4:    For an insoluble salt with one-to-one stoichiometry (such as $ZnS$, $TlCl$, or $PbCrO_4$) compare the exact and approximate solutions for $S$ in the presence of $C$ $M$ common ion with those for $[H_3O^+]$ in a solution of a strong acid (Chapter 2). By analogy to the latter case, sketch a graph of log $S$ vs. log $C$. The sharp decrease of $S$ as $C$ increases is generally referred to as the common-ion effect.

## Solubility of Salts with Basic Anions

A thunderstorm passes over a forest, pouring slightly acidic water (containing $H_2CO_3$ and its conjugate bases from the absorption of atmospheric $CO_2$, as well as traces of nitrogen acids formed by the action of lightning on atmospheric $N_2$ and $O_2$) over its limestone floor. Some of this water finds its way into a vertical crack in the rock. The limestone ($CaCO_3$) reacts with the acidity of the rainwater to form $HCO_3^-$:

$$CaCO_3(c) + H_3O^+ = Ca^{++} + HCO_3^- + H_2O$$

The solubility of the limestone is thus increased greatly over what one might expect on the basis of the $Q_{sp}$. When enough limestone has dissolved for the widening of the crack to be noticeable, a geologist calls it a *grike;* when it is wide enough to climb down into, a spelunker calls it a pit and hopes that it will turn out to be a full-sized cave. Whether they are large enough to satisfy tourists, spelunkers, or only bats, limestone caves are surely nature's most striking manifestation of solubility increased by the protonation of a basic anion.

For a quantitative understanding of this process and others like it, we must combine solubility and Brønsted acid-base equilibria. To be formal about it, we have in the case of $CaCO_3$ seven unknown molarities ($[Ca^{++}]$, $[CO_3^-]$, $[HCO_3^-]$, $[H_2CO_3]$, $[H_3O^+]$, $[OH^-]$, and $S$), and we need seven independent relationships among these. Although we will tackle the complete problem shortly, let us first solve a simpler one.

Very often in problems of this sort, $[H_3O^+]$ (and thus $[OH^-]$) is known beforehand, either through direct measurement or because the solution is buffered by some other equilibrium. In either case, the problem of calculating the solubility is much simpler. With $[H^+]$ and $[OH^-]$ out of the category of unknowns, we are down to five, and the following five relationships will solve the problem:

$$S = [Ca^{++}] \qquad (4\text{-}17)$$

since for every mole of $CaCO_3$ dissolved, one and only one mole of $Ca^{++}$ appears in the solution:

$$S = [CO_3^-] + [HCO_3^-] + [H_2CO_3]^1 \qquad (4\text{-}18)$$

$$Q_{sp} = [Ca^{++}][CO_3^-] \qquad (4\text{-}19)$$

$$Q_1 = \frac{[H^+][HCO_3^-]}{[H_2CO_3]} \qquad (4\text{-}20)$$

and

$$Q_2 = \frac{[H^+][CO_3^-]}{[HCO_3^-]} \qquad (4\text{-}21)$$

Equation 4-18 provides the best starting point:

$$S = \frac{Q_{sp}}{[Ca^{++}]} + \frac{[H^+][CO_3^-]}{Q_2} + \frac{[H^+]^2[CO_3^-]}{Q_1 Q_2}$$

$$S = \frac{Q_{sp}}{S} + \frac{Q_{sp}[H^+]}{Q_2 S} + \frac{Q_{sp}[H^+]^2}{Q_1 Q_2 S}$$

$$S^2 = Q_{sp}\left(1 + \frac{[H^+]}{Q_2} + \frac{[H^+]^2}{Q_1 Q_2}\right) \qquad (4\text{-}22)$$

---

[1] It has recently been recognized that $H_2CO_3$ is partly dehydrated to $CO_2$ even in aqueous solutions. The formerly established value of $Q_1$ ($4.6 \times 10^{-7}$) includes $[CO_2]_{aq}$ as part of the denominator. By using this value for $Q_1$, we will automatically include $CO_2$ in our equilibrium system and in our calculations.

Equation 4-22 allows a calculation of the solubility in a solution of known pH.

EXERCISE 4.5: Supply the algebra for the derivation of Eq. 4-22 from Eq. 4-18. Have all of the independent equations been used? Could you get there with fewer?

The expression in parentheses in Eq. 4-22 should look familiar, although we did not derive it in Chapter 2. It is the Brønsted equivalent to Eq. 3-4a, and the reciprocal of $\alpha_{CO_3^{--}}$. That is,

$$S^2 = \frac{Q_{sp}}{\alpha_{CO_3^{--}}} \qquad (4\text{-}23)$$

Equations 4-22 and 4-23 are really very sensible equations. They simply say that the solubility is different from $(Q_{sp})^{1/2}$ (the solution for an ordinary symmetrical salt in pure water) only to the extent that protonation makes $\alpha_{CO_3^{--}}$ different from unity. As $\alpha_{CO_3^{--}}$ decreases, solubility increases. $\alpha_{CO_3^{--}}$ will be essentially unity unless $[H^+]$ is of the order of or greater than $Q_2$.

Equation 4-22 has two nice graphical expressions. First, over a considerable range of acidity (i.e., $[H^+]$ less than $Q_1$), practically all the carbonate will be in the forms $CO_3^{--}$ and $HCO_3^-$, which makes the term $[H^+]^2/Q_1Q_2$ much smaller than the others in the parentheses. Equation 4-22 then has the simpler approximate form which is exact for a salt containing a monoprotic base:

$$S^2 = Q_{sp} + \frac{Q_{sp}}{Q_2}[H^+] \qquad (4\text{-}24^*)$$

This is a linear equation in $[H^+]$ with slope $Q_{sp}/Q_2$ and intercept $Q_{sp}$ and, as such, has been used as an experimental method of measuring $Q_{sp}$ and/or $Q_2$ in experiments in which solubility is determined as a function of known $[H^+]$. Figure 4-1 is a graph for $CaCO_3$ up to the point at which positive deviations due to the $[H^+]^2/Q_1Q_2$ term begin to be noticeable. Alternatively, Equation 4-22 may be graphed in logarithmic form (Fig. 4-2), in which case the relationship of $S$ to $\alpha_{CO_3^{--}}$ becomes obvious (compare Fig. 2-8). We will shortly find this logarithmic graph to be helpful.

**Solubility in pure water.**    Equation 4-22 is helpful if you know $[H^+]$, but it is useless otherwise. For example, the solubility of $CaCO_3$ in pure water still is related to the pH of the equilibrium solution, but the latter depends on the quantity of $CO_3^{--}$ which enters the solution; i.e., on the solubility. To get around this circularity, we must do a full-scale treatment based on Eqs. 4-17–4-22 and others, but before we plunge in like mathematicians, let us be sensible. We are contemplating a solution in which a diprotic base

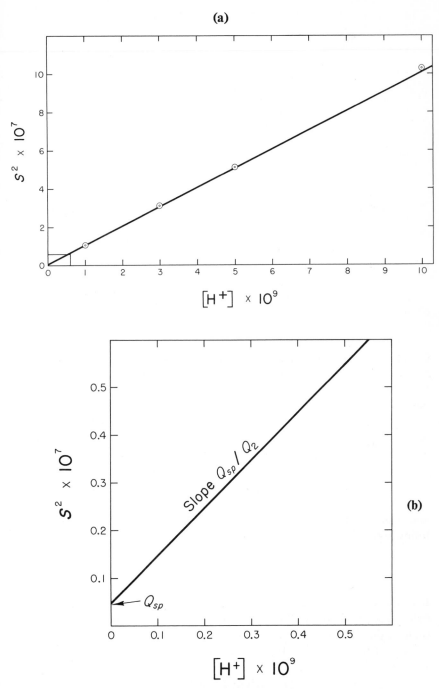

**Fig. 4-1.** (a) $S^2$ *vs.* [H$^+$] for CaCO$_3$ solubility in acidic solutions. The line is a graph of Eq. 4-24*, the circles are points calculated from Eq. 4-22. (b) Region near the intercept of Fig. 4-1a expanded to show intercept and limiting slope of $Q_{sp}/Q_2$ according to Eq. 4-24*.

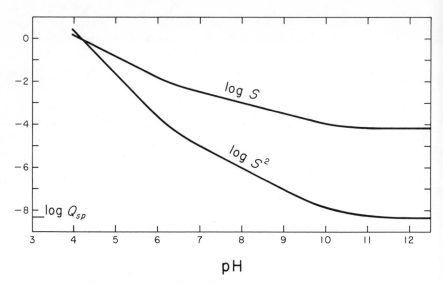

**Fig. 4-2.** Log $S^2$ and log $S$ for $CaCO_3$ as a function of pH. Compare Fig. 2-8 and Eqs. 4-22 and 4-23.

is dissolved in pure water. In our study of such systems in Chapter 2, we found that a solution of a polyprotic acid (or base) in water usually behaves as if the second and succeeding proton transfers did not occur. In the present example, this is to say that we may ignore $[H_2CO_3]$ in the mass balance (Eq. 4-18).

We now have 6 unknowns, and we have discarded a useful relationship ($Q_1$) along with $[H_2CO_3]$, so that we need two more independent equations. Since we now face the element of unknown pH, we have

$$Q_w = [H^+][OH^-] \qquad (4\text{-}25)$$

and a proton balance from the reference level $H_2O + CO_3^{--}$ gives us the following:

$$[H^+] + [HCO_3^-] = [OH^-] \qquad (4\text{-}26)$$

Because we have problems enough already, let us agree to treat the proton balance approximately. We will surely find that the solution is basic rather than acidic or even neutral, so we may say that $[H^+]$ will be very small compared to $[OH^-]$, which makes the proton balance

$$[HCO_3^-] = [OH^-] \qquad (4\text{-}27^*)$$

We may begin our development with Eq. 4-22 in the truncated form which results from dropping $[H_2CO_3]$ in the mass balance:

$$S^2 = Q_{sp}\left(1 + \frac{[H^+]}{Q_2}\right) \tag{4-28*}$$

where this time we do not know $[H^+]$. Whatever it is, though,

$$[H^+] = \frac{Q_w}{[OH^-]}$$

and from Eq. 4-27*,

$$[H^+] = \frac{Q_w}{[HCO_3^-]}$$

In view of the mass balance, this becomes

$$[H^+] = \frac{Q_w}{(S - [CO_3^{--}])}$$

or

$$[H^+] = \frac{Q_w}{S - (Q_{sp}/S)} = \frac{SQ_w}{S^2 - Q_{sp}} \tag{4-29*}$$

Insertion of Eq. 4-29* into Eq. 4-28* gives us an equation only in $S$ and known quantities:

$$S^2 = Q_{sp}\left(1 + \frac{SQ_w}{Q_2(S^2 - Q_{sp})}\right) \tag{4-30*}$$

or

$$Q_2S^4 - 2Q_2Q_{sp}S^2 - Q_{sp}Q_wS + Q_2Q_{sp}^2 = 0 \tag{4-31*}$$

This is a very recalcitrant equation. It will not factor completely, because $Q_w$ appears in only one term. Since the solution (for a general case) depends both on $Q_2$ and $Q_{sp}$, plotting a linear variable against assumed values of $S$ (which has only one solution for any system) has no particular advantage; worst of all, for the particular case of $CaCO_3$, the values of the constants are such that the terms containing $S^4$ and $S^2$ are just as important as the others, and so may not be discarded even as an approximation. Discarding the $S^4$ term leads to a predicted solubility *lower* than $Q_{sp}^{1/2}$, which is the minimum and only applies when protonation of the anion is negligible.

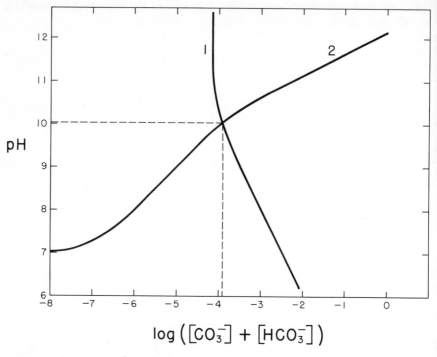

$$\log \left( \left[ CO_3^- \right] + \left[ HCO_3^- \right] \right)$$

**Fig. 4-3.** Line 1: Log solubility (Cf. Eq. 4-18 and Fig. 4-2) *vs.* pH for $CaCO_3$, assuming pH arbitrarily variable. Line 2: pH *vs.* C for the diprotic base $CO_3^{--}$, assuming C arbitrarily variable.

The intersection of these two lines meets both conditions, and thus represents the solubility of $CaCO_3$ in pure water.

A solution by successive approximations may be avoided by a graphical maneuver. We have a relationship for log $S$ as a function of pH (Fig. 4-2), and a graph of pH *vs.* log C for a weak base (the basic analog of Fig. 2-10). If these are plotted together, their intersection must represent the solution to this problem, since at that intersection, the solubility and acid-base equilibria are satisfied simultaneously. In Fig. 4-3, line 1 is a graph of log $S$ *vs.* pH calculated from Eq. 4-22 (compare Fig. 4-2), and line 2 is pH *vs.* $\log([CO_3^{--}] + [HCO_3^-])$; since $S = [CO_3^{--}] + [HCO_3^-]$, the coordinates are identical in meaning for the two lines. Note that their intersection occurs where both lines are curved, which accounts for our difficulty; no simple equation describes either the solubility or the pH. From the intersection at log $S = -3.91$, pH $= 10.04$, we deduce the values:

$$S = [Ca^{++}] = 1.23 \times 10^{-4}\ M$$

(the experimental value for the solubility of $CaCO_3$ in $CO_2$-free water is $1.3 \times 10^{-4}$ $M$):

$$[H^+] = 9.1 \times 10^{-11} \ M$$

$$[CO_3^{--}] = \frac{Q_{sp}}{[Ca^{++}]} = 3.8 \times 10^{-5} \ M$$

$$[HCO_3^-] = \frac{[H^+][CO_3^{--}]}{Q_2} = 8.1 \times 10^{-5} \ M$$

We may check these numbers through the mass balance:

$$[Ca^{++}] = [CO_3^{--}] + [HCO_3^-]$$
$$= (3.8 + 8.1) \times 10^{-5} \ M$$
$$= 1.19 \times 10^{-4} \ M$$

This compares to the value read from the graph to within 3 or 4%, which is average-to-good for values read from a logarithmic graph. These values may be refined by successive approximations to any desired accuracy, but they are already within the uncertainty in the various equilibrium quotients employed in the graphing.

This problem was worked out in detail more as an illustration of the complexities of simultaneous equilibria (and the power of graphical analysis to resolve them) than as a realistic problem. It is fortunately true that in most cases the value of at least one of the unknowns, such as the pH, is known *a priori*.

The solubility of metal sulfides in pure water is similar in behavior to that of carbonates, in that $S^{--}$ is quite a strong base ($pK_2$ of $H_2S$ is about $10^{-13}$); in fact, the effect is likely to be quite severe in most cases, because most metals form extremely insoluble salts with $S^{--}$

EXERCISE 4.6: What would be the effect of decreasing $Q_{sp}$ on a graph like Fig. 4-3? Does the effect of protonation on the solubility become more or less severe? How can you rationalize this trend?

As a result, any calculation of the solubility of sulfides that does not take into account protonation of $S^{--}$ is likely to be quite invalid. In fact, because metals that form insoluble sulfides also tend to form sulfide complexes (for example, $K_1$ for the $Ag^+$-$S^{--}$ complex is $10^{23.9}$), metal sulfides are perhaps the worst possible vehicle for the introductory teaching of equilibrium.

However, carbonates and sulfides are to some extent unrepresentative of the solubility of salts with basic anions, in that most common anions are much less basic than these two.

EXERCISE 4.7:   What happens to a graph like Fig. 4-3 when the anion is less basic than $CO_3^{--}$? (*Note:* Both lines change!) Sketch the analogous graph for calcium oxalate ($pQ_{sp} = 8.6$, $pQ_1 = 1$, $pQ_2 = 4$).

## Solubility with Simultaneous Equilibria Involving the Metal Ion

Having worked through the case of protonation of the anion of an insoluble salt, there is nothing really new to say about the effect of interaction of the cation with Lewis bases. Consider a salt MX such that the metal ion $M^+$ forms complexes either with $OH^-$ or with the Lewis base B. Then as a general starting point, if the anion $X^-$ undergoes no competing reactions,

$$S = [X^-]$$
$$S = [M^+] + [MOH] + [MOH_2^-] + \cdots + [MB] + [MB_2] + \cdots$$
$$[M^+] = S\alpha_0$$

and

$$Q_{sp} = S^2\alpha_0 \qquad\qquad (4\text{-}32)$$

Obtaining the value of $\alpha_0$ is simple if the stability constants of all complexes *and* the concentrations of $OH^-$ and B are known *a priori*, and difficult but possible if only the constants are known.

EXERCISE 4.8:   Rearrange Eq. 4-32 to an expression for log $S$ as a function of log $\alpha_0$, and, by reference to Fig. 3-2, sketch a graph of log $S$ for AgBr ($K_{sp} = 7.7 \times 10^{-13}$) in $Na_2S_2O_3$ solutions *vs.* log $[S_2O_3^{--}]$. It is this effect which allows "hypo" ($Na_2S_2O_3$) to render photographs insensitive to light by dissolving the unexposed AgBr emulsion.

An insoluble salt whose metal ion forms hydroxy-complexes alone and whose anion is nonbasic is mathematically analogous to $CaCO_3$; and equations analogous to 4-22 and 4-31* may be written with $[OH^-]$ replacing $[H^+]$, and stability quotients replacing acid dissociation quotients. In fact, except for the insoluble halides, examples of such salts are rather rare.

EXERCISE 4.9:   In view of the effect of anion-cation forces on solubility and the soft-hard correlations for the stability of complex ions, can you rationalize this last statement?

# Complexation of the Cation of the Salt by the Anion

A corollary of the answer to exercise 4.9 is that there are numerous instances of the formation of stable complexes between the anion of an insoluble salt and its cation. The silver salts are particularly prone to this behavior, and since they are such pedagogically prominent compounds, unsuspected complexation of $Ag^+$ by such anions as $Cl^-$, $SCN^-$, and the like gives the lie to such typical problems as the following:

What is the solubility of AgSCN in 0.1 $M$ KSCN? The $Q_{sp}$ is $1 \times 10^{-12}$.

An answer based solely on the $Q_{sp}$ would be $1 \times 10^{-11}$ $M$. However, because of the formation of $Ag^+$-$SCN^-$ complexes, the correct answer is quite different.

If we measure the solubility of AgSCN in the presence of excess $SCN^-$ by the sum of all forms containing $Ag^+$ (since the dissolution of AgSCN is the only source of $Ag^+$ in solution), we have[2]

$$S = [Ag^+] + [AgSCN] + [AgSCN_2^-] + [AgSCN_3^{-2}] + [AgSCN_4^{-3}] \quad (4\text{-}33)$$

In the presence of solid AgSCN, each of the terms on the right-hand side of Eq. 4-33 varies with $[SCN^-]$ in a particularly simple way as follows:

$$[Ag^+] = \frac{Q_{sp}}{[SCN^-]} \quad (4\text{-}34)$$

(Note that Eq. 4-34 is none the less true for the formation of complex ions; it simply does not describe all of the equilibria involved.)

$$[AgSCN] = Q_1[Ag^+][SCN^-]$$
$$= Q_1 Q_{sp} \quad (4\text{-}35)$$

$$[Ag(SCN)_2^-] = Q_2[AgSCN][SCN^-]$$
$$= Q_2 Q_1 Q_{sp}[SCN^-] \quad (4\text{-}36)$$

EXERCISE 4.10:   Note that we have terms in all powers of $[SCN^-]$ from $-1$ to $+1$ so far. What will be the terms for $[Ag(SCN)_3^{2-}]$ and $[Ag(SCN)_4^{3-}]$? Derive them formally to check your intuition.

The solubility, then, is represented by the series:

$$S = Q_{sp}[SCN^-]^{-1} + Q_1 Q_{sp} + Q_2 Q_1 Q_{sp}[SCN^-] + \cdots \quad (4\text{-}37)$$

---

[2][AgSCN] refers to the molarity of the soluble 1-1 complex, not the crystalline salt.

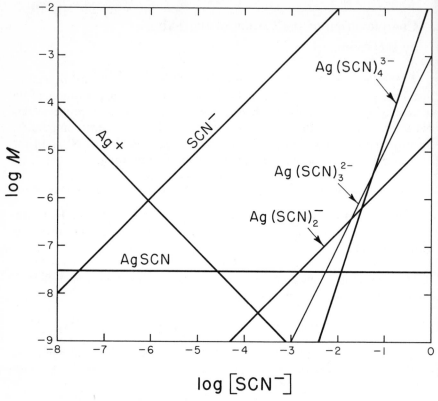

**Fig. 4-4.** Log molarity of all species in equilibrium with solid AgSCN as a function of log [SCN⁻]. The equilibrium quotients are valid for an ionic strength of 4 (Appendix 2).

The logarithm of any term in this sum varies in a simple way with log [SCN⁻]; for example,

$$\log [\text{AgSCN}_2^-] = \log (Q_2 Q_1 Q_{sp}) + \log [\text{SCN}^-] \qquad (4\text{-}38)$$

which suggests that a logarithmic plot should have a pleasing and useful appearance. Equation 4-38, on log molarity–log [SCN⁻] coordinates, would be a straight line with slope $+1$ and intercept $\log (Q_2 Q_1 Q_{sp})$; the other terms will have, respectively, slopes of $-1, 0, \ldots$, etc. Figure 4-4 is the resulting plot, with log [SCN⁻] (which depends on log [SCN⁻] in a particularly simple way) thrown in. From this graph we may see that, as for any other stepwise equilibrium, each of the species predominates over a definite range of the master variable (in this case, log [SCN⁻]). The variation of log $S$ with log [SCN⁻], then, is represented by the highest line on this graph at any particular value of log [SCN⁻] (Fig. 4-5). For example,

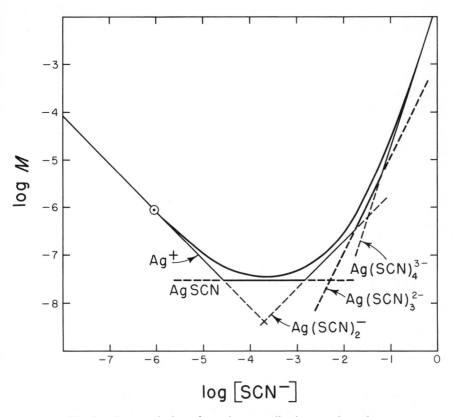

**Fig. 4-5.** Log molarity of species contributing to the solubility of AgSCN, as a function of log [SCN⁻]. Solid curved line: log $S$; solid straight lines: predominant species in the indicated range of log [SCN⁻]; dotted lines: species accounting for at least 10 % of $S$. Circled: log $S$ in pure water ([Ag⁺] = [SCN⁻]). Note that Ag⁺ predominates only below $2.5 \times 10^{-5}$ $M$ SCN⁻, and the $Q_{sp}$ permits an accurate calculation of $S$ only below $10^{-6} M$ SCN⁻.

below $10^{-5}$ $M$ SCN⁻ (but only there) the predominant term in Eq. 4-33 is [Ag⁺], and the solubility may be calculated with rough accuracy from the $Q_{sp}$ alone; but above 0.1 $M$ SCN⁻, $Ag(SCN)_4^{3-}$ predominates, and the solubility begins to be very large indeed.

To give a more correct answer to our original problem, the solubility of AgSCN in 0.1 $M$ KSCN is (from Fig. 4-5), $3 \times 10^{-5}$ $M$, and the dissolved silver is mostly in the form of $Ag(SCN)_4^{3-}$. This answer is greater than that calculated ignoring complex formation by a factor of 3 million.

It is worthwhile examining the crossing points of the lines in Fig. 4-4, since special conditions obtain there, and it is there that our approximation of representing log $S$ by the highest line is least valid. To settle this latter point first, we should recognize that, analogous to acid-base log graphs, log $S$ has a value log 2 higher than the ordinate of any crossing point; the smooth curve in Fig. 4-5 is drawn through these points. Also, at any crossing point, the condition that two concentrations are equal simplifies a stepwise stability quotient at that point. For example, at the $Ag^+$-AgSCN crossing, $[Ag^+] = [AgSCN]$; then

$$Q_1 = \frac{[AgSCN]}{[Ag^+][SCN^-]}$$

$$= \frac{1}{[SCN^-]}$$

and

$$\log [SCN^-] = -\log Q_1 \qquad (4\text{-}39)$$

EXERCISE 4.11:  The $Ag^+$ and $SCN^-$ lines cross at log $[SCN^-] = \frac{1}{2}(\log Q_{sp})$. Why?

These relationships make graphs such as Fig. 4-4 and Fig. 4-5 quite easy to construct. Beginning with the $M^+$-$X^-$ intersection at $\frac{1}{2}(\log Q_{sp})$, a line of slope $-1$ is drawn for log $[M^+]$. Where this line reaches log $[X^-] = pQ_1$, a line of slope zero crosses it, representing log $[MX]$, and at log $[X^-] = pQ_2$, this encounters the log $[MX_2]$ line, which has slope $+1$; and so on.

It might be useful at this point to introduce some nomenclature you are likely to encounter in the literature of equilibrium. For the equilibria listed below, the following equilibrium constants may be written:

$$MX(c) = M^+ + X^- \qquad (M^+)(X^-) = K_{s0}$$
$$MX(c) = MX(s) \qquad (MX)(s) = K_{s1}$$
$$MX(c) + X^- = MX_2^- \qquad (MX_2^-)/(X^-) = K_{s2}$$
$$MX(c) + 2X^- = MX_3^{--} \qquad (MX_3^{--})/(X^-)^2 = K_{s3}$$

etc.

Except for $K_{s0}$ (which we have called $K_{sp}$), these are products of what might be viewed as more fundamental quantities. For example, $K_{s1} = K_{sp}K_1$, etc. In general, it is preferable to list and discuss the more fundamental quantities $K_{sp}$ and $K_1$, $K_2$, etc., separately where possible. Occasionally, an experimental measurement of the solubility of a salt in the presence of excess anion will yield values of $K_{s1}$, $K_{s2}$, etc., without an independent value of $K_{sp}$. In that case, these constants cannot be resolved into their fundamental components, though $K_{s2}/K_{s1} = K_2$, etc.

The solubility quotient $Q_{s1}$ is sometimes referred to as the "intrinsic solubility" of a salt, on the grounds that its value is the molar concentration of a complex species of the same formula as the solid salt. Because the bonding of $M^+$ to $X^-$ in a 1-to-1 complex surely differs from the forces holding $M^+$ and $X^-$ in places in a crystal lattice, it is probably unwise to make much of the coincidence of chemical formulas between MX(c) and MX the soluble complex. Indeed, the notion of intrinsic solubility is a good example of an unfortunate human tendency to attribute physical significance to a mathematical artifact by giving it a suggestive name. Quantum mechanics offers several instances, of which resonance is perhaps the most famous.

## Suggestions for Further Reading

Butler, James N., *Ionic Equilibrium, A Mathematical Approach*. Reading, Mass.: Addison Wesley Publishing Co., 1964. Chapter 6.

Fleck, George M., *Equilibria In Solution*. New York: Holt, Rinehart and Winston, Inc., 1966. Chapter 2 takes a close look at precipitation titrations, in which solubility equilibrium is applied to analysis. Chapters 10 and 12 are careful considerations of particular systems.

Lewin, S., *The Solubility Product Principle*. London: Sir Isaac Pitman & Sons, Ltd., 1960.

Kolthoff, I. M., P. J. Elving, and E. B. Sandell, eds., *Treatise On Analytical Chemistry*. New York: Interscience Encyclopedia, 1959. Chapters 17 (by D. L. Leussing) and 19 (by J. F. Coetzee) include excellent discussions of solubility equilibria, and of the effect of atomic and molecular structure on equilibrium constants; Chapter 18 (by M. L. Salutsky) discusses such topics as the purity, particle size, and surface condition of ionic precipitates, and their effects on the physical and chemical behavior of the precipitates.

Ramette, R. W., "Meaningful Solubility Studies in Elementary Quantitative Analysis." *J. Chem. Ed.*, *33*, 610 (1956); "Solubility And Equilibria Of Silver Chloride." *J. Chem. Ed.*, *37*, 348 (1960). The application of solubility studies to the elucidation of solution equilibria is developed in these papers at a level appropriate for the reader of this book. The paper on AgCl will give some idea of the uncertainties with which all scientific investigation is fraught.

## Problems

*(Use data from Appendix 3 where necessary.)*

4.1. By what molar concentrations may total solubility (*S*) be measured in the following solubility equilibria? Give any necessary numerical coefficients.
   A. Bromine in pure water

B. $AgCl$ in pure water

C. $Ag_2CrO_4$ in pure water

D. $Ag_2CrO_4$ in a silver nitrate solution

E. $AgBr$ in an ammonia solution

F. $CaC_2O_4$ in an EDTA solution (high pH)

G. $CaC_2O_4$ in an EDTA solution (low pH)

4.2. The solubility of thallous iodide (TlI) in pure water is $2.5 \times 10^{-4}$ $M$. What is the $Q_{sp}$ of TlI?

4.3. What is the predominant $Ag^+$-$Cl^-$ complex in 0.1 $M$ $Cl^-$? What is the solubility of AgCl in 0.1 $F$ KCl? (You may want to draw a graph similar to Fig. 4-4.)

4.4. From the solubility product of AgBr and the formation quotients of the $Ag^+$-$Br^-$ complexes, determine the predominant form of silver in solution when pure water is saturated with AgBr. What is the total solubility, and to what extent do complex species contribute to it? (Hint: construct a graph similar to Fig. 4-4 and apply to it a bromide balance from the reference level AgBr.)

4.5. The $Q_{sp}$ of lead formate, $Pb(HCOO)_2$, is $2 \times 10^{-7}$.

A. Calculate the solubility of lead formate in pure water, ignoring protonation of the anion. Why is this simplification justifiable?

B. Calculate the solubility of lead formate in 0.1 $F$ NaHCOO; in 0.01 $F$ NaHCOO.

C. Calculate the solubility of lead formate in a well-buffered solution of pH 4.0.

D. Calculate the $[Pb^{++}]$ in a solution which is in solubility equilibrium with lead formate, and which contains 0.01 $M$ $H_3O^+$. This is not the same as the solubility of lead formate in 0.01 $F$ HNO$_3$. Why not? How could the latter solubility be calculated?

4.6. The pH of a solution containing 0.1 $F$ CaCl$_2$ is slowly raised. At what pH will Ca(OH)$_2$(c) first form?

4.7. The pH of an originally acidic solution containing 0.01 $F$ CaCl$_2$ and 0.1 $F$ oxalic acid is slowly raised. At what pH will calcium oxalate first precipitate? (Hint: construct a log distribution graph for 0.1 $F$ oxalic acid.)

4.8. Formulate a problem similar to 4.6 and 4.7 in which AgBr precipitates from an ammonia solution. Solve it.

4.9. The $Q_{sp}$ of $Sr_3(PO_4)_2$ is $1 \times 10^{-31}$.

A. What would be the solubility of strontium phosphate in pure water based only on this information?

B. What consideration does this calculation omit?

# 5 Distribution Equilibrium Across a Liquid Phase Boundary

"Remember the missing sugar? Well, it's turned up. Your fine friends dumped the whole bloody lot into the petrol. We're completely immobilized."

"Only one thing he can do — wash it. What size drums does your petrol come in?"

"Ten gallon."

"Tell him to pour out a couple of gallons and replace with water. Stir well. Let it stand for ten minutes and then syphon off the top seven gallons. It'll be as pure petrol as makes no difference."

"As easy as that!" I said incredulously. I thought of Hillcrest's taking half an hour to distill a cupful. "Are you sure, Mr. Mahler?"

— Alistair Maclean, *Night Without End.*

Mr. Mahler (a former research chemist who found himself in the midst of a spy thriller set in remotest Greenland at $-68°F$) was sure, because he knew the effect of polarity and hydrogen bonding on solubility. Since polar, hydrogen-bondable sugar is much more soluble in polar, hydrogen-bonding water than in nonpolar gasoline, virtually all the sugar can be removed from the sabotaged gasoline by giving it a large reservoir of water in which to dissolve.

In this chapter we will explore some of the applications of like-dissolves-like (see Chapter 4) when two immiscible solvents are placed in contact and a solute is allowed to come to equilibrium across the resulting phase boundary. Imagine, then, that a sample of gasoline contains some sugar (sucrose), and that some water is added to this and the mixture agitated gently to hasten equilibrium. The principal equilibrium established is

$$\text{sucrose (in water)} = \text{sucrose (in gasoline)} \tag{5-1}$$

and its equilibrium quotient is given the symbol $P$, for *partition coefficient*, or $D$, for distribution coefficient.

$$P^{o/w} = \frac{[\text{sucrose}]_o}{[\text{sucrose}]_w} \qquad (5\text{-}2)$$

where $o$ and $w$ refer, respectively, to the organic gasoline and to water. Because it will occasionally be convenient to think about the inverse of $P^{o/w}$ (which would be $P^{w/o}$), we will retain the superscript to be sure of which partition coefficient we mean at all times. We will take the symbol $o$ as a general one for the nonpolar phase since most, but not all, nonpolar liquids are organic substances.

Because the two liquids involved are necessarily very different in polarity, it is not difficult to predict qualitatively the value of $P^{o/w}$ for a given species if we know its polarity. For example, we should expect a nonpolar substance like $I_2$ to have $P^{o/w}$ values greater than unity, and this is indeed the case. With toluene ($C_6H_5CH_3$) as the organic solvent, $P^{o/w}$ for $I_2$ is 97.5 at 25°C; with $CS_2$ it is 600; and with amyl alcohol ($C_5H_{11}OH$) it is about 220.

> The fact that these values do not lie in the same order as the polarity of the organic solvents should warn us that a detailed prediction of the value of $P^{o/w}$ must take into account important second-order interactions of the solute with its solvent. That $I_2$ interacts with hydrogen-bonding solvents like water and the alcohols is indicated by its change of color from purple (as in the vapor phase and in non-hydrogen-bonding solvents) to brown.

There is, as we have been hinting, a relationship between the equilibrium distribution of a solute between two solvents and the ratio of the solubility of that solute in one solvent to its solubility in the other. Imagine a possible experiment in which excess solid $I_2$ is allowed to come to solubility equilibrium first with water and then with $CCl_4$. What will happen when these two saturated solutions are presented to each other? Nothing. We may summarize the relationships this way:

$$(5\text{-}3)$$

and by the very important principle: Two things in equilibrium with the same thing are in equilibrium with each other. To borrow the language of mathematicians, equilibrium is transitive. If this were not so, there would be a permanent net flow of $I_2$ molecules around equilibrium 5-3, and we could construct a little machine to get "useful" work out of the system without any input of energy, that is, a perpetual motion machine.

For the general case, then, we may conclude that

$$P^{o/w} = \frac{\text{(solubility in organic solvent)}}{\text{(solubility in water)}} \qquad (5\text{-}4^*)$$

There is one slight but sometimes important approximation involved in relationship 5-4* which accounts for the *. When two solvents are exposed to each other, a slight but significant quantity of each solvent dissolves in the other. That is, there are no perfectly immiscible solvents, so that the terms "water phase" and "organic phase" are really labels for the majority component in each phase. For example, when $CCl_4$ and water are shaken together at 25°C, equilibrium is reached when the water phase contains 0.001 mole fraction $CCl_4$ and 0.999 mole fraction $H_2O$, and the organic phase contains 0.001 mole fraction water and 0.999 mole fraction $CCl_4$. It can be worse than this; the mutual solubility of water and pentanol produces an organic phase containing nearly 0.34 mole fraction $H_2O$, and a water phase containing 0.001 mole fraction pentanol; and the solubility of diethyl ether in water is about 0.01 mole fraction. The significant point is that we will be considering not the pure solvents we have encountered up to now, but solvents whose properties are modified (perhaps only slightly) by the presence of traces of the other solvent. This fact makes Eq. 5-4* an approximation, since the solubility of a particular solute obviously depends on the composition of the solvent. The effect of solvent impurity should also be borne in mind when we consider the influence of other equilibria in the water phase. The values of such equilibrium quotients as $Q_a$'s, complexation quotients, and the like may be detectably different in these necessarily impure solvents from their values in purely aqueous solutions.

Some other partition coefficients are given in Table 5.1. Note the effect on $P^{o/w}$ when three polar, hydrogen-bonding OH groups are substituted for H on triethylamine (to form triethanolamine). A careful study of $P^{o/w}$ for sugar, gasoline, and water does not appear to have been made (Mr. Mahler was simply using the same reasoning we have to predict a rough value of $P^{o/w}$), but the datum for diethyl ether as a nonpolar phase should be indicative. If you worked problem 3.3, the comparison between $HgCl_2$ and KCl should not be surprising. The low value of $P^{o/w}$ for alanine compared to those for benzoic acid and triethylamine is something of an artifact. The substance is a zwitterion (problem 2.8) in water and neutral in nonpolar solvents, so that the single number $P^{o/w}$ is not a measure of a single equilibrium. The matter is further explored in problem 5.2.

TABLE 5.1.  SOME PARTITION COEFFICIENTS AT 25°C

| SUBSTANCE | ORGANIC SOLVENT | $P^{o/w}$ |
|---|---|---|
| $I_2$ | Toluene, $C_6H_5CH_3$ | 97.5 |
| $I_2$ | $CS_2$ | 600 |
| Benzoic acid, $C_6H_5COOH$ | Benzene, $C_6H_6$ | 2.1 |
| Alanine, $H_2N\overset{\overset{\text{H}}{\vert}}{C}\!\!-\!\!COOH$, $\underset{CH_3}{\vert}$ | Diethyl ether, $H_3CCH_2OCH_2CH_3$ | $1.4 \times 10^{-6}$ |
| Triethylamine, $(C_2H_5)_3N$ | Diethyl ether | 5.9 |
| Triethanolamine, $(HOCH_2CH_2)_3N$ | Diethyl ether | $1.1 \times 10^{-3}$ |
| $HgCl_2$ | Benzene | $8.4 \times 10^{-2}$ |
| $HgCl_2$ | Diethyl ether | 2.3 |
| $HgCl_2$ | $n$-Amyl alcohol, $C_5H_{11}OH$ | 1.8 |
| KCl | $n$-Amyl alcohol | $2.1 \times 10^{-3}$ |
| Sucrose, $C_{12}H_{22}O_{11}$ | Diethyl ether | $1.1 \times 10^{-6}$ |

# Effect of Competing Equilibria in the Aqueous Phase

We should recall from Chapter 4 that when a substance is capable of interactions (such as Brønsted or Lewis acid-base reactions) which change its charge or polarity there is an effect on its solubility. Since the distribution equilibrium is practically the ratio of the solubilities in the two solvents, any equilibrium that affects these solubilities will be reflected in the distribution. Again, consider $I_2$. Table 5.2 contains data (in modified form)

TABLE 5.2.  DISTRIBUTION EQUILIBRIUM DATA FOR $I_2$ BETWEEN $CCl_4$ AND AQUEOUS SOLUTIONS OF $BaI_2$

| $[I^-]$, $M$ | $C_w \times 10^{3\,a}$ | $C_o \times 10^{3\,a}$ | $C_w/C_o$ |
|---|---|---|---|
| 0.0424 | 3.953 | 7.82 | 0.505 |
| 0.0875 | 4.598 | 4.60 | 1.00 |
| 0.1294 | 8.051 | 5.48 | 1.47 |
| 0.2160 | 60.68 | 24.4 | 2.49 |

[a] $C$ = total formal concentration of iodine in the indicated phase. Adapted from J. N. Pearce and W. G. Eversole, *J. Phys. Chem.*, **28**, 245 (1924).

taken by investigators studying the distribution of $I_2$ between $CCl_4$ and aqueous $BaI_2$. The ratio of total iodine in the aqueous solution to that in the $CCl_4$ is far from constant; indeed it appears to increase in proportion to the $I^-$ concentration in the aqueous phase. The answer is, of course, that the interaction

$$I_2 + I^- = I_3^-; \quad Q_f = \frac{[I_3^-]}{[I_2][I^-]} \tag{5-5}$$

in the aqueous phase causes the solubility in that phase to increase relative to the solubility in $CCl_4$. We may define a quantity $E^{w/o}$ (the extraction coefficient) analogous to the quantity $S$ in solubility studies. $E^{w/o}$ is the equilibrium ratio of total formal concentrations of the solute in the two phases. In this case,

$$E^{w/o} = \frac{[I_2]_w + [I_3^-]}{[I_2]_o} \tag{5-6}$$

and we will employ analogous definitions for other systems. Since $[I_3^-] = Q_f[I_2]_w \cdot [I^-]$, we may rewrite Eq. 5-6 as

$$E^{w/o} = \frac{[I_2]_w + Q_f[I_2]_w \cdot [I^-]}{[I_2]_o} \tag{5-7}$$

In view of the definition of $P^{w/o}$ this is

$$E^{w/o} = P^{w/o} + P^{w/o}Q_f[I^-] \tag{5-8}$$

EXERCISE 5.1:   In the course of this derivation, the assumption was made that no $I^-$ entered the $CCl_4$ phase, and that no $I_3^-$ formed there. Can you justify this assumption?

EXERCISE 5.2:   Compare Eq. 5-8 with your result from exercise 4.1. What graphical display does Eq. 5-8 suggest? Try the data of Table 5.2 in such a graph.

Seeing that, at least in the case of $I_2$, $E^{w/o}$ depends on other equilibria in the same way that $S$ does, we may expect that much of what we have encountered in Chapter 4 will be directly applicable to the extraction coefficient $E^{w/o}$. The following discussions are meant more as an outline for your own thinking about these topics than as a complete exposition. We may begin to be grateful that, having reached this point, we may apply concepts first encountered in another context without re-deriving them each time.

**The $E^{w/o}$ of electrically neutral Bronsted acids and bases.**   Consider HX, which we will specify, since it is important this time, to have no electrical charge. Its conjugate base $X^-$ may be present in water, but will be too polar to have appreciable solubility in most nonpolar liquids. In that case, we may write for $E^{w/o}$ of HX:

$$E^{w/o} = \frac{[HX]_w + [X^-]}{[HX]_o} \tag{5-9}$$

At this point, we could go through a complete formal derivation of $E^{w/o}$ as a function of the pH,

> EXERCISE 5.3:   Do so, beginning with a substitution for $[X^-]$ in terms of $Q_a$, $[H^+]$, and $[HX]_w$, and continuing along the lines of the derivation of Eq. 5-8. How about a graph?

but by this time you should be able to do it yourself. Another approach, though, might be equally enlightening:

$$E^{w/o} = \frac{C_w}{C_o} \tag{5-10}$$

where $C_w = [HX]_w + [X^-]$, and $C_o = [HX]_o$. But $C_w$ may also be written as

$$C_w = \frac{[HX]_w}{\alpha_{HX}} \tag{5-11}$$

so that

$$E^{w/o} = \frac{[HX]_w}{[HX]_o \cdot \alpha_{HX}} = \frac{P^{w/o}}{\alpha_{HX}} \tag{5-12}$$

Let us again emphasize that since $X^-$ is assumed to exist only in the aqueous phase, the $\alpha_{HX}$ is intended to apply only to that phase.

> EXERCISE 5.4:   Suppose that you want to include the possibility of dissociation in the nonaqueous phase. Derive the modified form of Eq. 5-12 for that case.

Equation 5-12 in logarithmic form is

or

$$\left. \begin{aligned} \log E^{w/o} &= \log P^{w/o} - \log \alpha_{HX} \\[2mm] \log E^{o/w} &= \log P^{o/w} + \log \alpha_{HX} \end{aligned} \right\} \tag{5-13}$$

Equation 5-13 states that a graph of $\log E^{w/o}$ as a function of pH will resemble an upside-down logarithmic distribution graph of $\log \alpha_{HX}$ with a low-pH horizontal limit of $\log P^{w/o}$. This sort of graph is shown for benzoic acid in Fig. 5-1. Compare this graph with Fig. 2-5.

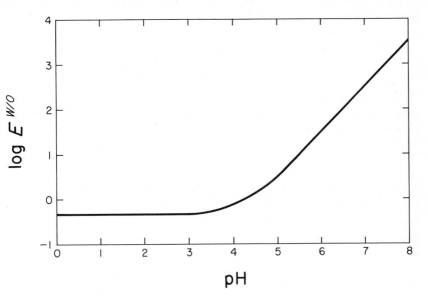

**Fig. 5-1.** Log $E^{w/o}$ *vs.* pH for benzoic acid in the solvents benzene and water. See Table 5.1 and Eq. 5-13.

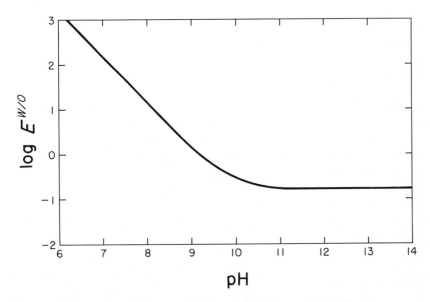

**Fig. 5-2.** Log $E^{w/o}$ *vs.* pH for triethylamine in the solvents diethyl ether and water. See Table 5.1.

Figure 5-2 is a graph of $\log E^{w/o}$ for the electrically neutral Brønsted base triethylamine (cf. Table 5.1).

EXERCISE 5.5:   Explain the shape of Fig. 5-2 and derive equations analogous to Eqs. 5-12 and 5-13 to back up your explanation.

$E^{w/o}$ **for complexes of metal ions.**    If we assume that a metal-Lewis base complex will have appreciable solubility in a nonpolar solvent only if it is electrically neutral, we may, by analogy to Eq. 5-13, immediately write,

$$\log E^{w/o} = \log P^{w/o} - \log \alpha_n \qquad (5\text{-}14)$$

where $n$ is the electrically neutral member of the series $M^{+x}$, $MX^{+x-y}$, etc., formed by a metal ion of charge $+x$ with a ligand of charge $-y$. Since we have already derived equations for the fraction of any member of such a series as a function of ligand concentration in Chapter 3, and seen that they may be graphed in the same way as the components of a Brønsted equilibrium, we may immediately sketch the graph of $\log E^{w/o}$ we expect to find.

EXERCISE 5.6:   The stepwise stability quotients of the $Hg^{++}$-$Cl^-$ series of complexes are: $Q_1 = 10^{6.74}$, $Q_2 = 10^{6.48}$, $Q_3 = 10^{0.85}$, and $Q_4 = 10^{1.00}$ at an ionic strength of 0.5. Sketch a graph of $\log E^{w/o}$ vs. $\log [Cl^-]$ for $Hg^{++}$ with water and diethyl ether as solvents. See Table 5.1; if you have worked problem 3.2, your work is almost done.

**Distribution equilibria of metal chelates.**    For analytical purposes it is frequently useful to form an electrically neutral chelate $MY_n$ which may be extracted into a nonpolar solvent. In this way M is separated from other metal ions which may be present in the aqueous phase, but it is not converted to a neutral species by complexation. Such equilibria share all the complications of ordinary metal-chelon equilibria (Chapter 3); in addition the electrically neutral protonated form $H_zY$ of the chelon $Y^{-z}$ is in most cases very soluble in nonpolar solvents. Thus, in addition to the ordinary pH dependence of the conditional stability quotient $Q'$, the very concentration $C_Y$ of the chelon is pH-dependent, since the extractable fraction of the chelon $\alpha_{H_zY}$ depends on pH. The resulting complex equilibria may be resolved into relatively tractable fractions, as the following simplified example shows.

Suppose we have a chelon $H_2Y$, with acid dissociation quotients $pQ_1 = 5$ and $pQ_2 = 11$, which forms a chelate with metal $M^{2+}$, of formula $MY$ and stability quotient $Q_{MY} = 10^{12}$. Let us examine the pH dependence of $E^{w/o}$ for the metal in a system containing 0.1 $F$ excess chelon. Suppose further that $P_{MY}^{w/o} = 0.01$ and $P_{H_2Y}^{w/o} = 0.2$. The following summarizes the situation:

$$HY^- \rightleftharpoons Y^{--} + H^+$$

| | | |
|---|---|---|
| Aqueous phase | $+$ $H^+$ | $+$ $M^{2+}$ |
| | $\Updownarrow$ | $\Updownarrow$ |
| | $H_2Y$ | $MY$ |
| | $\Updownarrow$ | $\Updownarrow$ |
| Organic phase | $H_2Y$ | $MY$ |

The experimentally determinable variables are the various equilibrium quotients (including $P$'s) and the total formal concentration of Y in excess of the metal ion, $C_Y$. Some of the chelon will be present in the aqueous phase $(C_Y')$, and some will be in the organic phase in the form of $H_2Y$. Thus

$$C_Y' = [H_2Y]_w + [HY^-]_w + [Y^{-2}]_w$$

and

$$C_Y = C_Y' + [H_2Y]_o \qquad (5\text{-}15)$$

The desired quantity is $E^{w/o}$, the extraction coefficient for the metal. If we use $C_M$ as defined in Chapter 3 for the formal concentration of the metal not complexed by the chelon, we find

$$E^{w/o} = \frac{[MY]_w + C_M}{[MY]_o} \qquad (5\text{-}16)$$

Ordinarily with excess chelon present $C_M$ would be very small compared to $[MY]_w$. However, with much of the chelon in the organic phase as $H_2Y(o)$ this may not be true. The first step in expanding Eq. 5-16 is to perform the indicated division:

$$E^{w/o} = P_{MY}^{w/o} + \frac{C_M}{[MY]_o} \qquad (5\text{-}17)$$

Now,

$$C_M = \frac{[MY]_w}{Q_{MY}' C_Y'} \qquad (5\text{-}18)$$

according to the definitions of $C'_Y$ and the conditional stability quotient $Q'_{MY}$ (Chapter 3). We may deduce $C'_Y$ from Eq. 5-15 and $P^{w/o}_{H_2Y}$:

$$C_Y = C'_Y + \frac{[H_2Y]_w}{P^{w/o}_{H_2Y}}$$

$$= C'_Y + \frac{C'_Y \alpha_2}{P^{w/o}_{H_2Y}} \tag{5-19}$$

where $\alpha_2$ is the fraction $[H_2Y]_w/C'_Y$. Factoring $C'_Y$ out of Eq. 5-19, we have

$$C'_Y = \frac{C_Y}{1 + \alpha_2/P^{w/o}_{H_2Y}} \tag{5-20}$$

Substituting into Eq. 5-18, we obtain

$$C_M = \frac{[MY]_w(1 + \alpha_2/P^{w/o}_{H_2Y})}{Q'_{MY}C_Y} \tag{5-21}$$

When this bundle is substituted for $C_M$ in Eq. 5-17 we find (recognizing that the ratio $[MY]_w/[MY]_o$ is $P^{w/o}_{MY}$ as always)

$$E^{w/o} = P^{w/o}_{MY} + \frac{P^{w/o}_{MY}(1 + \alpha_2/P^{w/o}_{H_2Y})}{Q'_{MY}C_Y} \tag{5-22}$$

Equation 5-22 looks messy and it is. Furthermore, it is the solution to a somewhat simplified system in that we have assumed a one-to-one complex between $M^{2+}$ and $Y^{2-}$. Complications such as protonation of $Y^{2-}$ and competing complexation of $M^{2+}$, which were enough to occupy a whole section of Chapter 3, are hidden in the symbol $Q'_{MY}$. Nevertheless, we are now equipped to handle such complications, and we may allow the elegant language of graphs to reveal the simplicity in the apparent complexity of Eq. 5-22. The first step is to expand Eq. 5-22 to three terms:

$$E^{w/o} = P^{w/o}_{MY} + \frac{P^{w/o}_{MY}}{Q'_{MY}C_Y} + \frac{P^{w/o}_{MY}\alpha_2/P^{w/o}_{H_2Y}}{Q'_{MY}C_Y} \tag{5-23}$$

The three terms on the right-hand side of Eq. 5-23 represent, respectively, $E^{w/o}$ in the absence of any complications, a correction term for the protonation of $Y^{2-}$, and a term representing the complicating factor of distribution of $H_2Y$ across the phase boundary.

EXERCISE 5.7:   Justify the statements in the last sentence to yourself. Which of the terms do you think should be pH-dependent? In what way?

Each of the terms varies in a simple, or at least familiar, way with pH. This will become more apparent when we take the logarithm of each term:

*Term 1:* $P_{MY}^{w/o}$ is a simple equilibrium quotient representing the equilibrium ratio of MY in the two solvents, and is thus independent of pH, as is its logarithm.

*Term 2:* The pH dependence here is in the variation of $Q'_{MY}$ with pH. Let us assume that there is no competing complexation of $M^{2+}$ by another base, so that protonation of $Y^{2-}$ is the only source of pH dependence. (If $\alpha_M$ as defined in Chapter 3 should be different from unity for some practical case, this problem may be handled along the lines of Chapter 3 without changing the outlines of the problem in which we are now immersed.) Then with the assumption that $\alpha_M = 1$, we may write $Q'_{MY}$

$$Q'_{MY} = Q_{MY}\alpha_Y \tag{5-24}$$

where, as before, $\alpha_Y = [Y^{2-}]/C'_Y$; or

$$\log Q'_{MY} = \log Q_{MY} + \log \alpha_Y \tag{5-25}$$

Term 2 then becomes

$$\log (\text{term 2}) = \log \frac{P_{MY}^{w/o}}{Q_{MY}C_Y} - \log \alpha_Y \tag{5-26}$$

Since the first term on the right-hand side of Eq. 5-26 is a constant, the log of term 2 has the shape of an inverted $\alpha_Y$ graph (cf. Fig. 5-1).

*Term 3:* In addition to the constants and the $\alpha_Y$ pH dependence, term 3 also features an $\alpha_2$ pH dependence. These two combine to make a surprisingly simple result:

$$\log (\text{term 3}) = \log \frac{P_{MY}^{w/o}\alpha_2}{Q'_{MY}C_Y P_{H_2Y}^{w/o}}$$
$$= \log \frac{P_{MY}^{w/o}}{Q_{MY}C_Y P_{H_2Y}^{w/o}} + \log \left(\frac{\alpha_2}{\alpha_Y}\right) \tag{5-27}$$

Now,

$$\frac{\alpha_2}{\alpha_Y} = \frac{[H_2Y]/C'_Y}{[Y^{2-}]/C'_Y}$$
$$= \frac{[H_2Y]}{[Y^{2-}]}$$
$$\frac{\alpha_2}{\alpha_Y} = \frac{[H^+]^2}{Q_1Q_2}$$
$$\log \left(\frac{\alpha_2}{\alpha_Y}\right) = \log \frac{1}{Q_1Q_2} - 2\,\text{pH} \tag{5-28}$$

When this is inserted in Eq. 5-27, we find

$$\log\,(\text{term 3}) = \log \frac{P_{MY}^{w/o}}{Q_{MY}C_Y P_{H_2Y}^{w/o} Q_1 Q_2} - 2\,\text{pH} \qquad (5\text{-}29)$$

The first term on the right-hand side of Eq. 5-29 is a constant, and the second will give a slope (on a graph *vs.* pH) of $-2$.

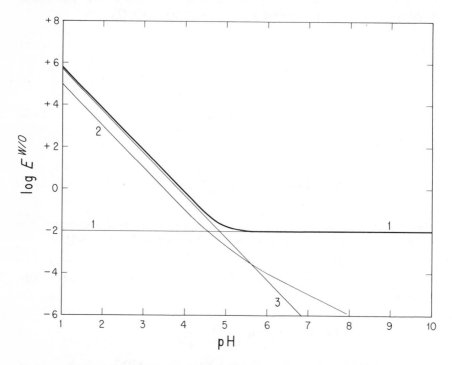

**Fig. 5-3.** Log $E^{w/o}$ *vs.* pH for $M^{2+}$ in the form of the chelate MY. See Eqs. 5-23–5-29. Line 1 corresponds to term 1; line 2 to term 2; line 3 to term 3; and the upper, heavy line to the sum of the three terms of Eq. 5-23.

The best way to see which of the three terms of Eq. 5-23 is important over a given range of pH is to graph each of them (or better, their logarithms) as a function of pH. The log $E^{w/o}$ will approximately follow the highest line on the graph. (This is the same trick we used in Fig. 4-5.) For the values of the constants and of $C_Y$ assumed in this problem, the resulting graph is shown in Fig. 5-3. The heavy line represents the log of the sum of the terms of Eq. 5-23; i.e., log $E^{w/o}$ as a function of pH. Note that at no time in this particular problem does term 2 predominate, though it is close to term 3 below a pH of 5. Note also that at pH $>$ 5, term 1 predominates; that is, there are "no" complications above a pH of 5 for this system. A graph like Fig. 5-3 is very helpful in planning experimental conditions for a chelate extraction of a metal ion.

## Applications of Distribution Equilibria

Equilibrium across a liquid-liquid phase boundary is not a common phenomenon in nature, unless you count the activities of mankind as natural. The justification for the elaboration of algebra in this chapter is not that it provides the key to some particular objective phenomenon like a limestone cave or küssmaul breathing, but that liquid-liquid partition is a very important tool for research and analysis in chemistry and biology.

**Equilibrium studies.** We have been assuming a knowledge of the equilibria and equilibrium quotients involved in a two-phase system and deducing the behavior of $E^{w/o}$ as a function of some variable such as pH or ligand concentration. However, the process may be reversed and experimental measurements of $E$ be interpreted in such a way as to deduce the values of equilibrium quotients. For example, the data in Table 5.2 were taken from a paper in which these results were used to prove the existence of the $I_3^-$ ion and to calculate the formation quotient $[I_3^-]/[I_2][I^-]$. Another example, involving an interaction in the organic phase, is explored in problem 5.1.

The disadvantage of such studies is one we have mentioned earlier: The solvents after equilibration are not pure; each contains traces of the other. If the equilibrium quotient being determined is sensitive to the nature of the solvent, the numerical value deduced will be of limited validity.

**Preparative chemistry.** Occasionally, the product of a chemical synthesis is enough unlike the starting materials in reactivity or polarity that the desired product may be isolated from the usual tangle of unreacted starting material and side products by means of a distribution equilibrium. For example, in the preparation of phenol ($C_6H_5OH$) the product, a weak acid with $pQ_a$ about 10, may be salvaged from the nonacidic, nonpolar reaction mixture by shaking the latter with an aqueous NaOH solution. $E^{w/o}$ for phenol is made quite large by the high pH of the aqueous solution, and it is effectively fished out of its nonpolar surroundings in the form of its conjugate base $C_6H_5O^-$. After the two phases are separated, phenol is recovered from the aqueous phase by lowering the pH, so that the less-polar and less-soluble acid form separates out. In other, less cut-and-dried situations it may not be possible to manipulate $E^{w/o}$ so dramatically, but a reaction product may be isolated from a mixture by means of chromatography, a technique based on two-phase distribution which we will explore later in this chapter.

**Separations involving a single equilibration.**    Very often in chemical work a mixture must be separated into its components. The usual reason for doing this is analysis; an originally unknown mixture is much more easily characterized if its components are separated. They are then identified and quantitatively measured one at a time. If the identity, but not the amounts, of two components in a mixture are known, then it is sometimes possible to manipulate their values of $E^{w/o}$ so that $E^{w/o}$ is large for one of the components and small for the other. In this case a single equilibration with two solvents results in a separation. As one of many possible examples, steel samples are much more easily analyzed for the minor components so important in determining their properties (such as the chromium in stainless steel) if part or all of the iron is separated from them before analysis. This can be achieved by oxidizing the iron to the $+3$ state and forming the chloro complexes $FeCl^{2+}$, $FeCl_2^+$, $FeCl_3$, $FeCl_4^-$, etc. When the solution contains roughly $6 F$ HCl, the anion $FeCl_4^-$ predominates and is extractable into diethyl ether as its conjugate acid $HFeCl_4$. Over $99\%$ of the Fe(III) is thereby removed from the aqueous solution (that is, $E_{Fe}^{w/o}$ is less than 0.01), while other metals in the sample remain in the aqueous phase. In problem 5.3 you will have an opportunity to explore a separation based on pH control.

**Effect of unequal phase volumes.**    Most often, two components of a mixture do not allow a clean separation in a single equilibration; there may be no conditions under which $E^{w/o}$ for one is large and for the other small. However, sometimes one component may be removed from a mixture even when its $E^{w/o}$ is only fairly small. For example, for triethylamine (T.E.A.) distributed between water and diethyl ether, $E^{w/o}$ has a minimum value at high pH of 0.17. However, if its removal from an aqueous solution is desired, one is more interested in the ratio

$$\frac{\text{(Moles T.E.A. in aqueous phase)}}{\text{(Moles T.E.A. in organic phase)}} = \frac{C_w V_w}{C_o V_o} \qquad (5\text{-}30)$$

$$= E^{w/o} \frac{V_w}{V_o} \qquad (5\text{-}31)$$

where $V_o$ is the volume of the organic phase and $V_w$ the volume of the aqueous phase. Equation 5-31 implies that a modest extraction coefficient may be made more favorable by a factor equal to the ratio of the volumes of the two phases used. By using a volume of diethyl ether 10 times that of the aqueous phase, the ratio of T.E.A. remaining in the aqueous phase to that transferred to the organic phase becomes 0.017 instead of 0.17.

EXERCISE 5.8:    Show that of the total number of moles of solute in both phases a fraction $E^{o/w}/(E^{o/w} + 1)$ winds up in the organic phase on equilibration, and a fraction $1/(E^{o/w} + 1)$ is in the aqueous phase if the volumes of the two phases are equal. How should this expression be altered for unequal volumes? What expression could be written using $E^{w/o}$ instead of $E^{o/w}$?

A ten-fold enhancement of any $E$ value is about all that may be reasonably achieved by using unequal volumes of the two phases, because if one phase becomes very small with respect to the other, it is difficult and time consuming to assure that all the large phase has had the opportunity to come to equilibrium with the smaller phase across the necessarily small phase boundary between the two.

A second technique for exhaustive extraction of a single component with only a modestly favorable $E$ is repetitive equilibration. For example, $E^{o/w}$ for As(III) (in the form of a neutral chloro complex) between 8 $F$ HCl and isopropyl ether is about 4. Suppose we equilibrate equal volumes of As(III)-containing 8 $F$ HCl and isopropyl ether. From the result of exercise 5.8, we may calculate that the fraction of As(III) removed from the aqueous phase is $4/(4 + 1) = 0.8$, and thus 0.2 (or 20%) of the original quantity remains, no matter how long we shake the two phases together. But now suppose we separate the two phases, add a second batch (again equal in volume to the aqueous phase) of pure, fresh isopropyl ether, and equilibrate again. In order to re-establish the distribution equilibrium, 80% of the remaining 20% [or 16% of the original As(III)] now enters the isopropyl ether, and 20% [only 4% of the original As(III)] remains in the 8 $F$ HCl.

EXERCISE 5.9:    Extrapolate this reasoning to predict how many equilibrations will be required to remove all but 0.1% of the original As(III) from the aqueous phase, and derive (or write down and check) equations for the fraction of extractable solute of extraction coefficient $E^{o/w}$ remaining in an aqueous phase after $n$ extractions with an equal volume of organic phase; do the same for unequal volumes.

# Countercurrent Distribution and Liquid Partition Chromatography

Any fundamental understanding of the functioning of biologically important macromolecules must always await their preparation or separation from natural sources in high purity, and the elucidation of their detailed structure. Industrial chemists often face the necessity of resolving complex mixtures of closely similar substances into their components, either for production purposes or for analysis of the purity of a product. Preparative chemists more often than not get the product that they want, along with several related compounds and unreacted starting material, in their flasks. In all of these problems, separations which formerly demanded weeks of tedious procedures may be accomplished quickly by means of various forms of chromatography, all of which depend on the repeated establishment of a distribution equilibrium between two phases. As an introduction to chromatography, we will consider first a closely related and conceptually simpler technique called *countercurrent distribution* which was developed by L. C. Craig in the early 1940's.

Imagine a long row of tubes containing a pure solvent which may be either polar or nonpolar and which we will call the stationary solvent. Into the first tube we place a single solute. In a practical case there would be a mixture of solutes to be resolved, but we will begin more simply. Now

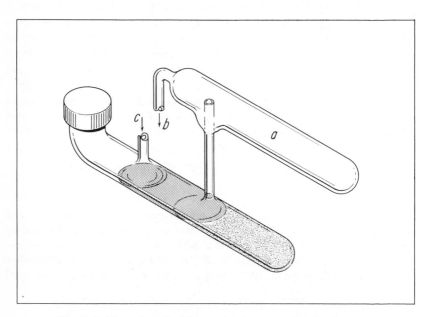

**Fig. 5-4.** The *n*th tube of a countercurrent distribution apparatus. During a transfer step the moving solvent goes into temporary storage in the holding tube *a*, and then enters tube number (*n* + 1) at *b* while moving solvent from the (*n* − 1)st tube's holding tube enters the *n*th tube at *c*. During an equilibration step the assembly of tubes is tipped backward to prevent any accidental sloshing into the holding tubes while it is rocked gently to hasten distribution equilibrium. Courtesy Pope Scientific, Inc. (adapted).

from a large reservoir we introduce into this same first tube a second solvent which we will call the moving solvent and which is immiscible with the stationary solvent. Tube 1 now contains the moving solvent, and the solute dissolved in the stationary solvent, with a phase boundary between the two solvents. The rest of the tubes still contain only pure stationary solvent. The solute now reaches an equilibrium distribution between the two solvents in the first tube according to the appropriate value of $E$. Because the distinction between the moving and stationary solvents is more important than the question of which of them is the aqueous phase, the extraction coefficient we want to use is

$$E^{m/s} = \frac{\text{(formal concentration of solute in moving phase)}}{\text{(formal concentration of solute in stationary phase)}}$$

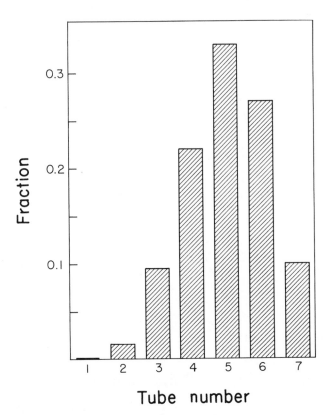

**Fig. 5-5.** Fraction of the total solute initially present in tube 1 after six equilibrations and transfers. Note that practically no solute remains in tube 1, and that the remainder has formed a concentration peak at tube 5. $E^{m/s} = 2.0$.

We now transfer the moving solvent (hence its name) by means of appropriate plumbing connecting the tubes (Fig. 5-4) from tube 1 to tube 2 and introduce a new batch of pure moving solvent into tube 1. Tube 1 now contains $1/(E^{m/s} + 1)$ of the original quantity of solute (assuming equal volumes of the two solvents in each tube, which is reasonable) and tube 2 contains $E^{m/s}/(E^{m/s} + 1)$ of it. All the other tubes still contain only stationary solvent, but their turn is coming. We now allow the solute in both tubes 1 and 2 to reach an equilibrium distribution between the two solvents in those tubes and again transfer each portion of moving solvent to the next tube down the line, introducing a fresh batch of moving solvent into tube 1. Clearly, if we continue this process many more times the concentration of solute in tube 1 will approach zero. The rather surprising behavior of the solute in the remainder of the tubes is perhaps best realized by referring to Table 5.3 and Fig. 5-5, which is a graph of the data of Table 5.3 for six equilibrations and transfers.

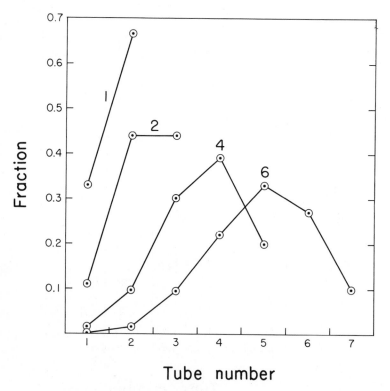

**Tube number**

**Fig. 5-6.** Fraction in each tube of the total solute after the indicated number of equilibrations and transfers. The circled points correspond to the tops of the bars in Fig. 5-5, and the connecting lines have no significance except to indicate more clearly the concentration peak.

**TABLE 5.3.**   COUNTERCURRENT DISTRIBUTION OF A SINGLE SUBSTANCE, $E^{m/s} = 2.0$.

| STAGE OF DISTRIBUTION | TUBE NUMBER | | | | | | |
|---|---|---|---|---|---|---|---|
| | 1 | 2 | 3 | 4 | 5 | 6 | 7 |
| Initial[a] | 0 | — | — | — | — | — | — |
| | 1.0 | — | — | — | — | — | — |
| Distribution 1 | 0.67 | — | — | — | — | — | — |
| | 0.33 | — | — | — | — | — | — |
| Transfer 1 | 0 | 0.67 | — | — | — | — | — |
| | 0.33 | 0 | — | — | — | — | — |
| T-1 Totals | 0.33 | 0.67 | — | — | — | — | — |
| Distribution 2 | 0.22 | 0.44 | — | — | — | — | — |
| | 0.11 | 0.22 | — | — | — | — | — |
| Transfer 2 | 0 | 0.22 | 0.44 | — | — | — | — |
| | 0.11 | 0.22 | 0 | — | — | — | — |
| T-2 Totals | 0.11 | 0.44 | 0.44 | — | — | — | — |
| Distribution 3 | 0.07 | 0.29 | 0.29 | — | — | — | — |
| | 0.04 | 0.15 | 0.15 | — | — | — | — |
| Transfer 3 | 0 | 0.07 | 0.29 | 0.29 | — | — | — |
| | 0.04 | 0.15 | 0.15 | 0 | — | — | — |
| T-3 Totals | 0.04 | 0.22 | 0.44 | 0.29 | — | — | — |
| Distribution 4 | 0.025 | 0.15 | 0.29 | 0.20 | — | — | — |
| | 0.013 | 0.07 | 0.15 | 0.1 | — | — | — |
| Transfer 4 | 0 | 0.025 | 0.15 | 0.29 | 0.20 | — | — |
| | 0.013 | 0.07 | 0.15 | 0.1 | 0 | — | — |
| T-4 Totals | 0.013 | 0.095 | 0.30 | 0.39 | 0.20 | — | — |
| Distribution 5 | 0.01 | 0.063 | 0.20 | 0.27 | 0.14 | — | — |
| | — | 0.032 | 0.10 | 0.13 | 0.07 | — | — |
| Transfer 5 | 0 | 0.01 | 0.063 | 0.20 | 0.27 | 0.14 | — |
| | — | 0.032 | 0.10 | 0.13 | 0.07 | 0 | — |
| T-5 Totals | — | 0.04 | 0.16 | 0.33 | 0.34 | 0.14 | — |
| Distribution 6 | — | 0.027 | 0.11 | 0.22 | 0.23 | 0.10 | — |
| | — | 0.013 | 0.05 | 0.11 | 0.11 | 0.04 | — |
| Transfer 6 | — | — | 0.027 | 0.11 | 0.22 | 0.23 | 0.10 |
| | — | 0.013 | 0.05 | 0.11 | 0.11 | 0.04 | 0 |
| T-6 Totals | — | 0.013 | 0.077 | 0.22 | 0.33 | 0.27 | 0.10 |

[a]In each pair the upper number gives the fraction of the initial quantity of solute currently present in the *moving* phase in that tube, and the lower number gives the fraction present in the stationary phase. In a distribution step, the total quantity of solute in both phases of a tube is distributed in a 2-to-1 ratio in accordance with the value of $E^{m/s}$. In each transfer step, the moving phase of each tube is transferred to the next tube in the line.

Table 5.3 and Fig. 5-5 show two very important features: first, instead of smearing out uniformly over the tubes, the solute tends to remain together so that a "peak" of concentration is observed. Second, this peak moves down the row of tubes. It is true that there is some inevitable spreading of the solute over a range of tubes. In this example after three equilibrations and transfers over 95% of the solute is concentrated in the three tubes at the center of the peak; after five, only 83% is. Figure 5-6 shows the tube contents after various numbers of equilibrations and transfers. Note that the peak becomes better defined as it spreads out and moves, because its rate of spreading is slower than the rate at which the moving solvent travels.

Though it is not obvious from this consideration of a single solute, the peak moves in the direction of the moving solvent at a rate which depends on its $E^{m/s}$. In Fig. 5-7 we see the distribution after 1, 5, and 10 transfers for two substances, one with $E^{m/s} = 1.0$, and the other with $E^{m/s} = 2.0$. The peaks have begun a process of separation which, though still incomplete

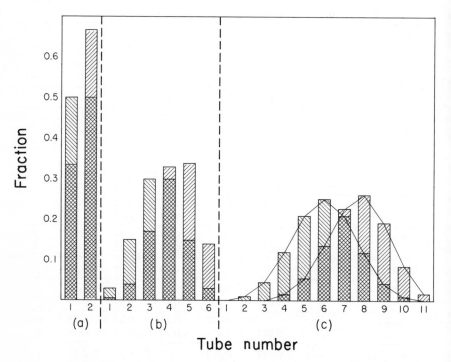

**Fig. 5-7.** Countercurrent distribution of a mixture of two substances, one [//////] with $E^{m/s} = 2.0$, the other [\\\\\\] with $E^{m/s} = 1.0$. The fraction of each initial quantity per tube is shown after one (a), five (b), and ten (c) stages of equilibration and transfer. Only in (c) are the two peaks well enough resolved for connecting lines to be helpful.

at this stage, will continue with repeated equilibrations and transfers until any desired degree of separation is obtained, while obviously a one-step separation of these two substances is impossible. A single countercurrent apparatus may comprise hundreds of tubes, and it is always possible to recycle the moving solvent from the last tube to the first, so that it is not difficult to use thousands of equilibrations and transfers to accomplish separations. It is, of course, necessary to have some method of keeping track of solute peaks as they move along the apparatus. The usual technique is to monitor the outflow of moving solvent as it leaves the last tube, using a characteristic of the solute which is distinct from that of the moving solvent, such as visible or ultraviolet absorption, rotation of polarized light, pH, or refractive index, to detect solute peaks as they appear.

Countercurrent distribution is especially useful in organic and biological chemistry, since molecules very little different from each other in polarity (and thus in $E^{m/s}$) may be separated when no other method of separation is possible. In fact, complex mixtures whose components are mostly unknown may be sorted out on the basis of polarity as a sort of surveying technique, and interesting peaks isolated for further study. A typical example is the investigation of the purple meadow rue (see Suggestions for Further Reading).

**Liquid partition chromatography.**   A technique similar in principle to countercurrent distribution, but somewhat more efficient and easier to use, is liquid partition chromatography. The same equilibrium of a solute between a moving and a stationary solvent is used, but the two solvents are in continuous contact across a phase boundary between the stationary phase, which is adsorbed onto the surface of a finely divided solid support, and a moving phase which percolates through this finely-divided, solvent-coated solid. The operation is usually carried out in a vertical column, with the moving solvent added at the top, trickling through the bed of stationary solvent, and exiting through a stopcock at the bottom (Fig. 5-8). In the course of running slowly over the very finely-divided stationary phase, the moving phase effectively equilibrates with the stationary many times per centimeter of column height, and a rather small column (a typical one might be one meter long) may accomplish separations which would require thousands of steps in a countercurrent apparatus.

The rate at which a particular solute moves in the direction of the moving solvent is characterized by its $R_F$, defined by

$$R_F = \frac{\text{(location of solute peak)}}{\text{(location of moving solvent front)}} \qquad (5\text{-}32)$$

For example, in Fig. 5-7, after 10 equilibrations and transfers, the $R_F$ of substance A is $6/11 = 0.55$. Strictly speaking, the $R_F$ is not a constant

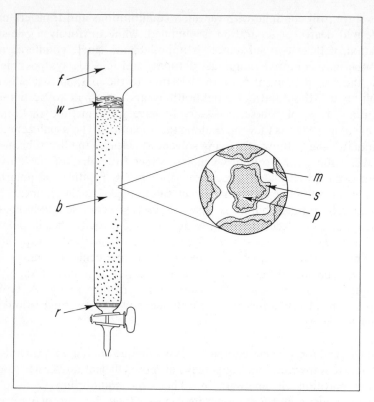

**Fig. 5-8.** A typical liquid partition chromatography tube. $f$: funnel top for ease in adding the mixture to be separated and the moving solvent; $w$: glass wool to prevent addition of moving solvent from disturbing the contents of the column; $b$: bed of solid particles coated with stationary solvent (see blown-up cross section); $r$: sintered glass disc to prevent the bed of stationary solvent-coated particles from exiting through the stopcock; $m$: moving solvent: $s$: stationary solvent; $p$: particle of the solid which supports the stationary solvent.

(as you may verify in Fig. 5-6 or 5-7), but after a relatively small number of equilibrations approaches the value $E^{m/s}/(E^{m/s} + 1)$.

EXERCISE 5.10:   Where have we seen this fraction before? The above expression follows if we think of $E^{m/s}$ as the ratio of the average times that a solute molecule spends in the moving and stationary solvents. Why is such an assumption plausible?

You may verify this relationship in the special case of $E^{m/s} = 1$ by noting that after $n$ equilibrations and transfers the solvent front is in tube number

($n + 1$) and the peak is in tube number $n/2$. As $n$ becomes very large, the ratio of these two numbers is clearly 0.5, which is $1/(1 + 1)$.

Even a fairly complete discussion of the many applications of countercurrent distribution and liquid partition chromatography would fill a book larger than this one. Let one further example suffice: a series of structurally similar carboxylic acids including acetic, fumaric, oxalic (see Chapter 2) and others may be separated by partition chromatography with an aqueous sulfuric acid stationary phase and a butanol-chloroform mixture as moving phase. The author once devoted a summer to a similar separation of these acids which were present as by-products in artificial cellulose gum. The purpose was a complete analysis which would convince the Food and Drug Administration that the stuff was safe as an additive to ice cream. It makes cheap formulations seem "creamier." Equilibrium, like atomic energy, is ethically neutral.

## Other Types of Chromatography

If an equilibrium can be established between a solute in a moving solvent and a stationary phase of any description, a chromatographic separation can be based on this equilibrium. An annotated list of methods may give you some idea of the variety of equilibria which can be exploited, and thus the variety of mixtures which can be resolved. The Suggestions for Further Reading at the end of this chapter will provide elaboration.

**Adsorption chromatography.** The moving solvent percolates through an uncoated solid adsorbent such as aluminum oxide. The equilibrium established for solute $s$ is

$$s \text{ (in moving solvent)} = s \text{ (on solid surface)} \qquad (5\text{-}33)$$

The classical example is Tswett's resolution of leaf juice into its component chlorophylls and carotenes in 1903. It was this separation of colored substances that gave the general technique of chromatography its name.

**Ion-exchange chromatography.** If electrically charged functional groups are covalently bonded to a cross-linked polymer, the counter-charges necessary for the structure's stability may be present as relatively simple ions, held in the resin structure by electrostatic forces alone. The entire structure is usually made sufficiently open and porous that when the resin is immersed in water, the liquid enters the structure and, as we have seen, diminishes the electrostatic force holding the simple ions to the charged polymer. These ions may now be exchanged for others originally present in solution in the water. A cation exchanger utilizes a polymer containing negative functional groups such as $-COO^-$ or $-SO_3^-$. The ion-exchange equilibrium for cations $c_1^+$ and $c_2^+$ on resin R is

$$c_1^+(aq) + R^-c_2^+ = c_2^+(aq) + R^-c_1^+ \qquad (5\text{-}34)$$

A typical resin may contain several millimoles of charge per gram. The equilibrium quotient for Eq. 5-34 depends mainly on the force between the resin binding site and the ions, and this force depends on all of the factors influencing Lewis acid-base reactivities. Thus, a mixture of even slightly different ions may be separated chromatographically with the ion-exchange resin in the role of the stationary phase. An auxiliary complexing agent may be used to change the reactivities, or indeed the sign of the charge, of some of the ions in the mixture. Any aqueous equilibrium affecting the properties of the exchangeable ions may thus be used to enhance their differences and their ease of separation. One of the first applications of the technique was the resolution of mixtures of the lanthanide ions. These ions are chemically so similar that their separation was formerly achieved through painfully slow fractional crystallizations.

An application of interest to the biologically inclined is the resolution of mixtures of amino acids and peptide groups. This process exploits not only the inherent differences in the molecules but also the dependence of their charge on the pH of the solution (see problem 2.10). It appears that for complex species adsorptive as well as purely electrical interactions occur between the amino acid and the ion exchanger. This complicates the interpretation, but not the utility of the method. A recent example is the determination of the amino acid sequences in harbor seal, porpoise, and whale myoglobin. Pure myoglobin was isolated from tissue extract, broken into fragments with a digestive enzyme, and the separated fragments completely characterized by a commercial amino acid analyzing instrument. All three stages of separation were achieved by ion-exchange-adsorption chromatography, employing control of the pH of the moving solvent.

*Paper chromatography.* This technique establishes an equilibrium between a moving solute (usually, but not always, nonaqueous) and the surface of filter paper, across which the solvent moves by capillarity. The exact nature of the stationary phase and the condition of the solute in it were not clear for some time, but it now appears that adsorbed water interacts with the paper to produce a hydrated but solid-like surface.

*Gas chromatography (vapor-phase chromatography).* The moving phase is a stream of a carrier gas, and the stationary phase is a liquid held on a finely divided solid support. Since the equilibrium involved is

$$s \text{ (vapor)} = s \text{ (stationary solvent)} \qquad (5\text{-}35)$$

separations may be made on the basis of the polarity of the stationary solvent and on the basis of the boiling point of the pure solute. An important variable is the temperature at which the separation is made, since this factor strongly influences the position of equilibrium 5-35. Because very high temperatures are sometimes used, it is important that the stationary phase have a very low vapor pressure. A typical stationary solvent might be a silicone oil or a polar, high-molecular weight substance like dibutyl phthalate.

Detection of the solute peaks as they emerge from the end of the chromatography column presents an unusual analytical problem, since they are present as dilute vapors in a stream of carrier gas. The standard method of detection involves automatic measurement and recording of the thermal conductivity of the effluent gas mixture, but other more sensitive and/or more specific methods have been developed. A recent example involves the use of an automatically recording hydrogenation apparatus which detects only compounds (mainly unsaturated hydrocarbons) that react with molecular hydrogen. The most elegant detector I have encountered was a group of male red-banded leaf roller moths. The effluent stream of a gas chromatograph was passed through their cage, and their general level of activity served as a detector of unparalleled economy, sensitivity, and specificity for the sex attractant pheromone of the female red-banded leaf roller.

An economically and ecologically important use of gas chromatography has been the detection and quantitative measurement of pesticide residues in foodstuffs, wild animals, and people.

## Suggestions for Further Reading

Giddings, J. C., "Physico-Chemical Basis of Chromatography." *Journal of Chemical Education, 44,* 704–709 (December 1967). The level of much of this discussion may exceed your present powers, but it is introduced and summarized by qualitative and gracefully phrased extensions of the ideas presented in this chapter. The current capabilities and future promise of chromatography in physical and biochemistry are surveyed by one who has contributed greatly to the present state of the art.

Heftmann, E. (ed.), *Chromatography*, 2nd Edition. New York: Reinhold Publishing Corp., 1967. A collection of contributions which will serve as a thoroughgoing text for the really interested student.

Inczedy, J., *Analytical Applications of Ion Exchangers.* New York: Pergamon Press, 1966. Included in this scholarly work is some discussion of ion exchange chromatography. It will be more useful as a fundamental introduction to the theory and practice of ion exchange itself.

Laitinen, H. A., *Chemical Analysis.* New York: McGraw-Hill Book Co., Inc., 1960. Chapter 25 proceeds from distribution equilibria through countercurrent distribution to chromatography in a fairly thorough and rigorous way. The discussion should be accessible to the reader at this point, though the author's symbols are different from ours.

Morris, C. J. O. R., and P. Morris, *Separation Methods in Biochemistry*. London: Sir Isaac Pitman and Sons, Ltd., 1963. A professional, but not too abstruse, treatment of the use of distribution equilibria and chromatography in achieving separations of biologically important substances, and in isolating particular substances from plant and animal tissue.

"Plants Supply Promising Antitumor Agents." *Chemical and Engineering News*, December 12, 1966, pp. 64ff. A popular account of the use of partition chromatography and countercurrent distribution to isolate products of biological and medical importance from such plants as the purple meadow rue. The cover of this issue of the magazine features a dramatic photograph of a countercurrent distribution apparatus.

Although it is not "suggested for further reading" because of its terse and professional level of presentation and its focus on other matters, the àrticle "Comparison of Myoglobin From Harbor Seal, Porpoise and Sperm Whale" [K. D. Hapner, R. A. Bradshaw, C. R. Hartzell, and F. R. N. Gurd, *Journal of Biological Chemistry*, *243*, 683–689 (1968)] contains the isolation of myoglobin and two-stage separation of its component amino acids referred to in this chapter.

# Problems

5.1. The following data describe the distribution equilibrium for formic acid between water ($w$) and benzene ($o$) at 25°C.

| Experiment | $C_w, F$ | $C_o, F$ |
|:---:|:---:|:---:|
| 1 | 2.574 | 0.00568 |
| 2 | 4.675 | 0.0123 |
| 3 | 6.265 | 0.0208 |
| 4 | 7.400 | 0.0265 |
| 5 | 8.242 | 0.0325 |
| 6 | 9.047 | 0.0378 |

A. Calculate $E^{o/w}$ for each experiment.

B. The variation in $E^{o/w}$ with the total quantity of formic acid in the system has been attributed to the dimerization of formic acid in the nonpolar phase (see Chapter 1), according to the scheme

$$HCOOH \text{ (aq)} = HCOOH(\text{benzene})$$
$$\|$$
$$(HCOOH)_2(\text{benzene})$$

Is the trend in $E^{o/w}$ consistent with this scheme? Why (not)?

C. Show that if the formic acid dimerizes in benzene, $E^{o/w}$ must be expressed as

$$E^{o/w} = \frac{[\text{HCOOH}]_o + 2[(\text{HCOOH})_2]_o}{[\text{HCOOH}]_w}$$

D. Show that the expression derived in C implies the equation

$$E^{o/w} = P^{o/w} + 2Q_d(P^{o/w})^2[\text{HCOOH}]_w$$

where $Q_d$ is the dimerization equilibrium quotient.

E. What graphical analysis does the equation in D suggest? Try the data in such a graph, and deduce values of $P^{o/w}$ and $Q_d$. Is the $P^{o/w}$ reasonable on the basis of the polarity and hydrogen-bonding capabilities of formic acid? How does the value of $Q_d$ compare with that given in Chapter 1?

F. What if someone suggested to you that dissociation of the formic acid in the aqueous phase is responsible for the nonconstancy of $E^{w/o}$? Is the trend in $E^{o/w}$ consistent with this scheme? Why (not)?

G. In fact, a simple experimental precaution was taken to ensure that practically none of the formic acid in the water phase was in the form of $\text{HCOO}^-$. Can you suggest what this precaution might be?

5.2. The partition coefficient for an amino acid such as that for alanine listed in Table 5.1 may be regarded formally as an over-all equilibrium quotient for a two-step process:

$$
\begin{array}{l}
\quad\quad \text{H} \quad\quad\quad\quad\quad\quad\quad \text{H} \\
\text{R---C---COOH } (o) = \text{R---C---COOH } (w) \\
\quad\quad \text{NH}_2 \quad\quad\quad\quad\quad\quad\quad \text{NH}_2
\end{array}
$$

$$
\begin{array}{l}
\quad\quad \text{H} \quad\quad\quad\quad\quad\quad\quad \text{H} \\
\text{R---C---COOH } (w) = \text{R---C---COO}^- (w) \\
\quad\quad \text{NH}_2 \quad\quad\quad\quad\quad\quad\quad \text{NH}_3 \\
\quad\quad\quad\quad\quad\quad\quad\quad\quad\quad\quad\quad + \\
\hline
\quad\quad \text{H} \quad\quad\quad\quad\quad\quad\quad \text{H} \\
\text{R---C---COOH } (o) = \text{R---C---COO}^- (w) \\
\quad\quad \text{NH}_2 \quad\quad\quad\quad\quad\quad\quad \text{NH}_3 \\
\quad\quad\quad\quad\quad\quad\quad\quad\quad\quad\quad\quad +
\end{array}
$$

A. With this simplification, write a definition for $E^{w/o}$ for an amino acid, taking into account the equilibria

$$
\begin{array}{l}
\quad\quad \text{H} \quad\quad\quad\quad\quad\quad\quad\quad\quad \text{H} \\
\text{R---C---COO}^- + \text{H}_2\text{O} = \text{R---C---COO}^- + \text{H}_3\text{O}^+ \\
\quad\quad \text{NH}_3 \quad\quad\quad\quad\quad\quad\quad\quad \text{NH}_2 \\
\quad\quad + 
\end{array}
$$

and

$$
\begin{array}{l}
\quad\quad \text{H} \quad\quad\quad\quad\quad\quad\quad\quad \text{H} \\
\text{R---C---COO}^- + \text{H}_2\text{O} = \text{R---C---COOH} + \text{OH}^- \\
\quad\quad \text{NH}_3 \quad\quad\quad\quad\quad\quad\quad\quad \text{NH}_3 \\
\quad\quad + \quad\quad\quad\quad\quad\quad\quad\quad\quad\quad +
\end{array}
$$

B. By comparison with Figs. 2-8, 5-1, and 5-2, sketch a graph of log $E^{w/o}$ *vs.* pH for the amphiprotic amino acid alanine.

5.3. Indicate how two weak bases with roughly equal, and small, values of $P^{w/o}$, but different $Q_b$'s, could be separated by means of a distribution equilibrium. (A graph showing log $E^{w/o}$ *vs.* pH for both bases will be helpful.)

5.4. A. In view of the relationship between $R_F$ and $E^{m/s}$ (p. 134), what is the limiting value of log $R_F$ when $E^{m/s}$ is very large? To what does log $R_F$ reduce when $E^{m/s}$ is small compared to 1?

B. For a base with $P^{o/w}$ small compared to 1, sketch a graph of log $R_F$ *vs.* pH of the aqueous phase for partition chromatography of that base employing an organic moving phase and an aqueous stationary phase.

5.5. What is the relationship between the order in which the separated components of a mixture emerge from a chromatography column and their $R_F$'s?

5.6. In adsorption chromatography using a nonpolar moving phase and a powdered sugar stationary phase, how would you expect a solute's $R_F$ to correlate with

A. its solubility in water?

B. its solubility in gasoline?

C. its Lewis basicity?

D. its ability to form hydrogen bonds?

5.7. In gas chromatography the $R_F$ is difficult to measure precisely, and a parameter called the "retention time," which depends on the carrier gas flow rate, is used as a working measure of the $R_F$. The retention time is the time between the injection of a sample at the head of the chromatography column and its appearance at the detector.

A. How (qualitatively) is retention time related to $R_F$?

B. How would you expect a substance's retention time to correlate with

(1) its solubility in the stationary phase?

(2) its boiling point?

(3) the temperature of the chromatography column?

(4) its polarity, assuming a nonpolar stationary phase?

# 6 Oxidation-Reduction Equilibria and Electrochemical Cells

## Oxidation-Reduction Reactions: A Review

The student who reads this book is supposed to have taken at least one general chemistry course. Yet it may be that like many students of chemistry and their teachers he approaches electron-transfer reactions uneasily. These reactions lend themselves to the creation of terms and definitions below the clear and simple surface of which lurk darkness and confusion. Fuzzy ambiguities about which element in an ion or molecule gains or loses electrons are swept under a rug of arbitrary conventions; definitions recalled from physics, such as those of anode and cathode, turn out to be limited or inappropriate; authors and teachers make distinctions (e.g., between the "American" and "European" sign conventions), the necessity and meaning of which elude the most generous and open-minded listener.

The difficulties in part reflect the complexity of the reactions. As electrons are the stuff of chemical bonds, their gain or loss may cause radical rearrangements of the atoms involved. Some oxidations are as "simple" as

$$Fe^{++} \rightarrow Fe^{3+} + e^- \qquad (6\text{-}1)$$

but it takes some insight to see how many electrons are lost, and from where, when

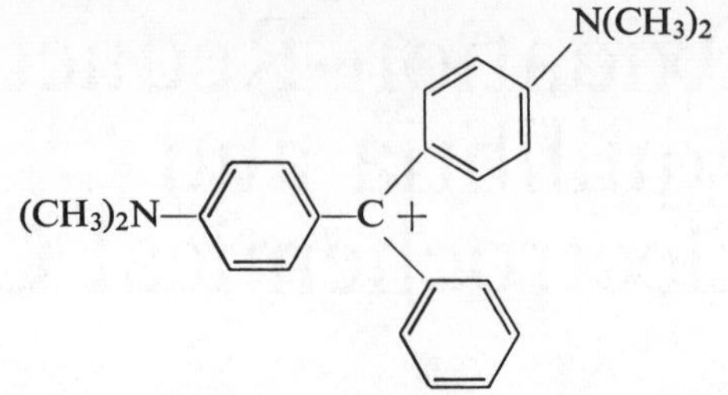

becomes

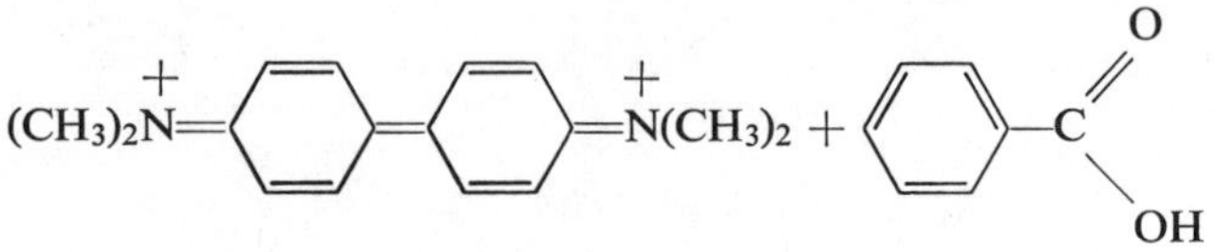

Likewise, you should have no difficulty in writing down the conjugate acid of the Brønsted base hydrazine, $H_2NNH_2$, but you might be at a loss to give the "oxidized form" of this reducing agent even if, out of several possibilities, a two-electron oxidation is specified. Thus, it may be useful to review some basic concepts and definitions.

**Oxidation and reduction:**   Oxidation is the loss of one or more electrons, and reduction is the gain of one or more electrons, by a chemical species. In some cases it is possible to describe precisely what species gains electrons, and what species loses them, as in

$$Cl_2(g) + 2K(g) \rightarrow 2KCl\,(c) \qquad\qquad (6\text{-}2)$$

In the course of the above reaction gaseous potassium atoms (the stable form of potassium at low pressure above 760°C) clearly lose one electron each in becoming $K^+$ ions in the KCl crystal lattice. They are thereby oxidized. Chlorine atoms, which have an octet through sharing in the molecule Cl:Cl, each have an octet of their own in the chloride ion. This is a relatively unambiguous gain of electrons, and the chlorine is thus reduced.

Let us consider another example:

$$Mn(c) + 8H_2O \rightarrow H_2(g) + 2OH^- + Mn(H_2O)_6^{++} \qquad\qquad (6\text{-}3)$$

An orthodox explication of this reaction would say that the Mn was oxidized, and the hydrogen in water was reduced. This analysis, however, cannot be advanced as a realistic description of the changes involved. Wherein is a hydrogen atom in H:H richer in electrons than one in H:O:H? Only in that the sharing of electrons in $H_2O$ is not uniform, but is polarized toward the O. On this basis, it would be more realistic to say that in Eq. 6-3 the oxygen atom is reduced, because in water it shares four electrons and has four of its own, whereas in $OH^-$ it shares two and has six of its own. Next consider the Mn atom. In the metal, it shares its two 4s electrons with every other Mn atom in a metallic bond. In $Mn(H_2O)_6^{++}$, each water molecule donates a pair of electrons to a covalent bond with the manganese ion. The resulting environment is surely not much poorer in electron density than that of a metallic crystal. If any atoms have lost electrons in the transaction, it is the oxygen atoms coordinated to the manganese, since they now share a pair of electrons which were previously their own (compare the conversion of $2Cl^-$ to $Cl_2$).

The oxidation-number system, with which you should be familiar, was devised to minimize this confusion. In some situations, the use of oxidation numbers provides a useful shorthand classification. Thus, we may consider $Fe_2O_3$, $Fe(CN)_6^{3-}$, and $Fe(H_2O)_6^{3+}$ to be compounds of $+3$ iron, even though the electronic environment of the iron may be quite different from one compound to the other. But it is not good to place too much credence in oxidation numbers as a description of the actual distribution of electrons in a compound. For example, the ion $ClO_4^-$ cannot realistically be described as an aggregate of $+7$ chlorine and $-2$ oxygen ions.

**Half-reactions:**   The ambiguities of oxidation-reduction lore need not, for the purposes of this book, concern us greatly. Any given oxidation-reduction reaction occurring in solution may be broken into half-reactions; for example,

$$H_2O_2 + 3I^- + 2H^+ \rightarrow I_3^- + 2H_2O \qquad (6\text{-}4)$$

which we can imagine to be the sum of the half-reactions

$$H_2O_2 + 2H^+ + 2e^- \rightarrow 2H_2O \qquad (6\text{-}4a)$$

and

$$3I^- \rightarrow I_3^- + 2e^- \qquad (6\text{-}4b)$$

Let us agree to remember that the electrons occurring in a half-reaction do not represent independent solutes which happen to cancel out in the addition of the half-reactions. Except in rare cases, electrons do not appear in solution as an independent solute, but are transferred from species to species in the course of a reactive collision. Electrons should never appear in an over-all oxidation-reduction equation.

Half-reactions as a device for systematizing oxidation-reduction reactions may claim a kind of realism which oxidation numbers cannot: a particular reaction may literally be split into half-reactions by carrying out the reaction

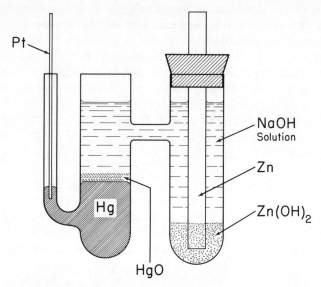

**Fig. 6-1.** A typical electrochemical cell. The platinum wire on the left is provided for convenient electrical contact to the mercury electrode. The cell potential is the difference in voltage of two pieces of the same metal (in practice, the leads to the voltage measuring device) attached to the platinum and to the zinc.

in an electrochemical cell. For the remainder of this chapter, we will consider how the properties of such cells may be understood on the basis of the chemical equilibria present in them.

## Electrodes and Electrochemical Equilibria

A typical electrochemical cell (Fig. 6-1) consists of two electrically conducting phases (nearly always metal) called electrodes, in contact with a solution. To prevent a direct reaction between the reactants at the two electrodes the cell may be divided into two half-cell compartments. Each half-cell contains one of the electrodes, and the solutions in them are allowed to make the necessary electrical contact in a way (such as through a glass-wool plug, or sintered glass disk) which minimizes mixing. The heart of the half-cell is the electrode-solution interface, at which a redox half-reaction equilibrium

$$a\text{A} + b\text{B} + \cdots + n\text{e}^- = c\text{C} + d\text{D} + \cdots \qquad (6\text{-}5)$$

is established. The electrons in Eq. 6-5 are mobile electrons within the electrode material. They have a variable activity which is dependent on the position of equilibrium 6-5, and which may be detected as a variable electrical potential on the electrode.

For scholarly discussions it is useful to distinguish between the electrical potential within the electrode where the mobile electrons are and the electrical potential outside, but very near, the electrode, which is the only potential accessible to measurement. Also, if the two electrodes are made of different substances, the difference in their potentials is measured as the difference in potential between two pieces of the same conducting substance (usually copper) attached to them. In practice, the internal wiring of any voltage measuring device serves this purpose. Neither of these considerations need concern us, since we are interested in interpreting observable, external potentials.

The physical forms in which electrical cells may occur vary greatly from such obvious examples as flashlight batteries to less obvious ones such as rust pits on an impure iron surface. Aside from the gross morphology described above, half-cells may differ in the types of substances in electrochemical equilibrium at the electrode surfaces, and it is useful, though not fundamentally necessary, to categorize them.

**(1) Metal-metal ion electrodes:** The half-reaction has the general form

$$M^{n+} + ne^- = M(c) \qquad (6\text{-}6)$$

Any metal and ion stable in the presence of the solvent may form this kind

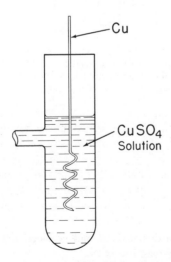

**Fig. 6-2.** A copper-cupric sulfate half-cell. The truncated arm on the left is connected to another half-cell in practice.

of electrode. Thus, silver-silver ion, copper-cupric ion, etc., form well-behaved half-cells (see Fig. 6-2). For obvious reasons, active metals such as those of groups IA and IIA do not form stable half-cells, nor do such metals as Zn in acid solution.

EXERCISE 6.1:   Why not?

**(2) Metal-metal salt electrodes:**   These are really the same as the above type. The equilibrium is

$$M_xY_y \text{ (c)} + nx\,e^- = xM \text{ (c)} + yY^{-nx/y} \qquad (6\text{-}7)$$

for a metal in the oxidation state $+n$ in the slightly soluble salt $M_xY_y$ (see Fig. 6-3). Examples are Ag-AgBr and Hg-Hg$_2$Cl$_2$.

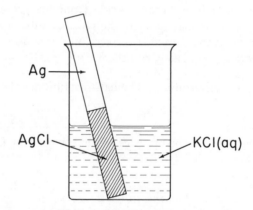

**Fig. 6-3.** A silver-silver chloride half-cell. The shaded area is a coating of AgCl on the Ag surface. A half-cell constructed in this way would be connected to the other half of the cell by means of a salt bridge (see Fig. 6-6b).

EXERCISE 6.2:   The rather complicated superscript on the Y in Eq. 6-7 follows from the stoichiometry of the insoluble salt. Verify it with a few examples.

The only difference between half-cells of the above types is that in the latter the concentration of the $M^{n+}$ ion is controlled by the solubility equilibrium of the salt and the concentration of the anion present.

**(3) Electrodes with both oxidized and reduced forms in solution:**   The general form of the half-reaction is

$$\text{ox} + ne^- = \text{red} \qquad (6\text{-}8)$$

where ox and red are the oxidized and reduced forms, respectively, of some substance, and are present in solution. These electrodes are sometimes called redox electrodes. Of the multitude of examples, we may illustrate with

$$Fe^{3+} + e^- = Fe^{2+}$$
$$I_2 + 2e^- = 2I^-$$

and

$$\underset{\text{quinone}}{C_6H_4O_2} + 2H^+ + 2e^- = \underset{\text{hydroquinone}}{C_6H_4(OH)_2} \qquad (6\text{-}9)$$

See Fig. 6-4. Here the electrons involved in the half-reaction do not have an obvious home, since all reactants and products are in solution. In this sort of half-cell, a metal electrode must be supplied, at the surface of which the reactants and products in solution and electrons in the metal are at equilibrium. Since it is clearly desirable that the metal not react chemically with any of the components of the half-cell, metals difficult to oxidize (i.e., the noble metals) are chosen for this role. The most commonly used metal is platinum, but gold, palladium, mercury, and even more reactive metals are used on occasion.

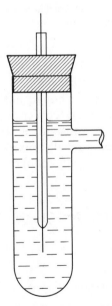

Fig. 6-4. A "redox" half-cell. A platinum or other noble metal electrode is in contact with a solution containing oxidized and reduced forms of a solute; for example, $Sn^{4+}$ and $Sn^{++}$, or $Cr_2O_7^-$ and $Cr^{3+}$, etc.

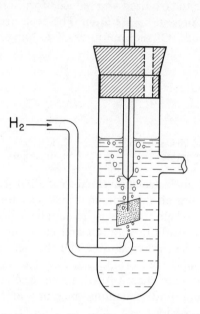

**Fig. 6-5.** A hydrogen half-cell. To insure equilibrium among $H_2$, $H^+$ and electrons in the platinum electrode, an electrode with a large surface is often used, and the surface is roughened by electrodepositing platinum in finely divided form ("platinum black").

**(4) Gas electrodes:** When either red or ox in a type 3 half-cell is a gas, some modifications in structure (though not in principles of operation) are required. To be sure that the concentration of the volatile gaseous component of the half-cell is constant, a stream of gas containing the reactant (usually pure, but occasionally mixed with other gases) is bubbled through the half-cell in the vicinity of the electrode surface (Fig. 6-5). Examples are

$$2H^+ + 2e^- = H_2 \text{ (in solution)} \qquad (6\text{-}10)$$

and

$$Cl_2 \text{ (in solution)} + 2e^- = 2Cl^- \qquad (6\text{-}11)$$

In both examples the dissolved gas is in solubility equilibrium with the gas phase.

## Liquid Junctions and Salt Bridges

Consider a cell composed of a silver-silver bromide half-cell and a ferric ion–ferrous ion half-cell (Fig. 6-6). The solutions in the two half-cells must be kept separate, since $Fe^{3+}$ is capable of oxidizing silver metal to silver bromide. In the cell of Fig. 6-6, a sintered glass disk allows electrical contact between the two solutions without appreciable mixing. Such a point of contact between solutions of dissimilar ions is called a liquid junction. Its importance for our discussion lies in the fact that an electrical potential called the liquid junction potential always arises at a liquid junction. The source of this potential is unequal transport of electrical charge across the junction by migration of the ions of the two solutions. Its magnitude is difficult to measure and, except in a few cases, to calculate. Methods for approximate calculation of liquid junction potentials exist, but are beyond the scope of this book. Since this potential is in series with the two electrodes, it is measured as part of the cell voltage and introduces unavoidable ambiguity into the measured potential of any cell with a liquid junction. As we will see, not all cells have a liquid junction; in particular, the one shown in Fig. 6-1 does not.

Fortunately, liquid junction potentials may be suppressed, although not eliminated entirely, by the use of a *salt bridge*. This is a concentrated inert salt solution which is inserted between the two half-cells, with a liquid junction at each end (Fig. 6-6b). Almost all the transport of charge is accounted for by the ions of the concentrated inert salt. If the salt is chosen properly ($KCl$ and $NH_4NO_3$ are frequently used), the cation and anion will transport nearly equal quantities of charge, and the liquid junction potential will be small, and may be neglected for most measurements.

## Schematic Representations of Cells

The kind of artwork which has decorated our discussion to this point is unnecessarily elaborate. In order to streamline their discourse, electrochemists have evolved a schematic method for representing cells, and we will use it in the remainder of this chapter. For example, the cell of Fig. 6-6 can be written as

$$Ag \mid AgBr, Br^-(C_{Br^-}) \mid Fe^{++}(C_{Fe^{++}}), Fe^{3+}(C_{Fe^{3+}}) \mid Pt \qquad (6\text{-}12)$$

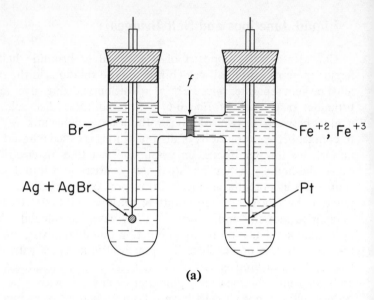

(a)

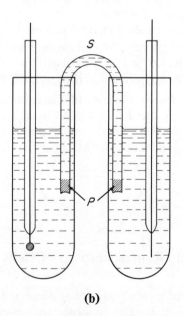

(b)

Fig. 6-6. (a) A typical cell with liquid junction. Contact
between the dissimilar solutions is made at a fritted glass
disk (*f*). (b) The same cell as (a), but with the half-cells
connected by a salt bridge (*S*) containing $NH_4NO_3$, which
makes contact with the half-cell solutions at two porous
plugs (*P*) which form liquid junctions.

The vertical solid lines represent electrode surfaces at which the respective half-reactions are at equilibrium and, between two solutions, a liquid junction. Any relevant information about the concentrations of the solutes in the half-cells is given with the symbol of the solute. A salt bridge in the cell is represented by a double vertical line between the solutions, unless the composition of the salt bridge is given, in which case single vertical lines surround it. For example, the cell of Fig. 6-6b can be represented either as

$$Ag \mid AgBr, Br^- \mid NH_4NO_3 \mid Fe^{++}, Fe^{3+} \mid Pt \qquad (6\text{-}13)$$

or simply as

$$Ag \mid AgBr, Br^- \parallel Fe^{++}, Fe^{3+} \mid Pt \qquad (6\text{-}14)$$

Information about the cell reaction is contained in the schematic representation; we will return to this topic.

## Reference Electrodes

The absolute electrical potential of a single electrode is not a measurable quantity. Fortunately, it is not of interest, since random static charges the cell and its electrodes may acquire render the absolute potential of any single part of it meaningless. A more useful quantity might be the potential of one electrode with respect to its half-cell solution, but this is also inaccessible. Any probe by which one would determine the potential of the solution must itself be electrically conducting and thus becomes a second electrode. We are thus driven to measure the *relative* potentials (that is, the difference in potential) of two electrodes and to make what we can of the information. Fortunately, we can make quite a bit.

Whenever one quantity must be measured in relation to another, it is helpful to agree on an arbitrary standard with which all others are compared. Thus, the absolute location of Geneva, New York is not measurable or even definable; but we may place it quite accurately relative to the locations of Greenwich, England and the equator by giving the coordinates 76°59′ west longitude, 42°52′ north latitude. As a working reference for anyone familiar with the local geography, it may be helpful to add that Geneva lies about 45 miles southeast of Rochester, New York. In exactly the same way we may define an arbitrary reference electrode against which the potentials of all others may be measured, and it may be useful to use various convenient working reference electrodes whose potentials relative to the arbitrary standard are known to most chemists. (It may comfort you to reflect that the familiar surface of the earth is two-dimensional and curved, whereas the scale of electrode potentials is one-dimensional and, as far as anyone knows, straight). We will return to a discussion of electrodes actually used as references after we consider the influence of concentrations of half-cell components on the potential.

## The Nernst Equation

As a focus for discussion, consider the cell

$$Pt, H_2 \mid H^+ \parallel Cu^{++} \mid Cu \qquad (6\text{-}15)$$

The voltage of the above cell (the difference in the potentials of the copper and platinum electrodes) is found to be 0.34 volt, with the copper positive relative to the platinum, when the activities of $H^+$, $H_2$, and $Cu^{++}$ are each unity. We now consider the importance of specifying the values of the activities of all species involved in the half-cell equilibria.

Suppose that the half-reaction

$$Cu^{++} + 2e^- = Cu \qquad (6\text{-}16)$$

is at equilibrium at $(Cu^{++}) = 1$. This equilibrium is characterized by a definite electron activity in the copper, and thus by a definite potential on the copper; relative to that on the platinum electrode, the potential is the above-mentioned $+0.34$ V. Now suppose that $(Cu^{++})$ is increased, for example by adding more $Cu^{++}$ to the half-cell solution. Common sense, in the form of Le Chatelier's principle, indicates that equilibrium 6-16 must shift to the right, as would any other equilibrium. This shift will cause a decrease in the activity of electrons on the copper electrode, which will be manifested as a change in the potential of the electrode.

EXERCISE 6.3: Since electrons are, by convention, negatively charged, in which direction will the potential of the copper electrode change when $(Cu^{++})$ is increased in the half-cell solution? When it is decreased? Write the half-reaction which is at equilibrium in the hydrogen half-cell, and predict the effects of changes in $(H^+)$ and $(H_2)$ on the potential of the platinum electrode.

In order to convert these qualitative conclusions into something numerical and thus generally useful, we must return to our consideration of the cell as a whole. Since for this copper-hydrogen cell the copper electrode is more positive than the hydrogen electrode, we could cause a current of electrons to flow from one electrode to the other by connecting them with a wire (Fig. 6-7). Since electrons flow from negative to positive points in a circuit, electrons would leave the platinum, flow through the wire, and

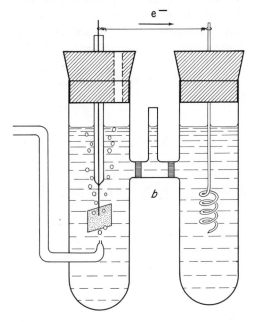

**Fig. 6-7.** The cell

$$Pt. \; H_2|H^+||Cu^{++}|Cu$$

short-circuited by means of a wire connecting the electrodes. Electrons flow from negative to positive (in this cell from Pt to Cu) through the wire.

enter the copper.[1] But if this occurs, the half-reaction equilibria at both electrode surfaces will be upset: Equation 6-16 will shift to the right as more electrons are supplied to the copper, and the other half-reaction

$$2H^+ + 2e^- = H_2 \qquad (6\text{-}17)$$

will shift to the left as electrons are removed. The sum of these two half-reaction shifts is the complete reaction

$$H_2 + Cu^{++} \rightarrow 2H^+ + Cu \qquad (6\text{-}18)$$

---

[1]The electrode at which electrons enter the cell is defined as the *cathode:* that at which electrons leave the cell is defined as the *anode.* We will not have great use for these terms, but you will encounter them in other electrochemical discussions. Note that these definitions say nothing about the sign of the charge on the cathode and anode. In this, as in all spontaneously operating cells, the cathode is positive, the anode negative. It follows from the definitions that reduction always occurs at the cathode, oxidation at the anode.

There is thus associated with any electrochemical cell a spontaneous cell reaction which takes place if the electrodes are connected with an external conductor. The chemical energy of the cell reaction is the source of the electrical energy appearing as an electrical potential difference at the electrodes.

Reaction 6-18 cannot proceed to the right unless the proper quotient of activities[2]

$$J = \frac{(H^+)^2}{(H_2)(Cu^{++})} \tag{6-19}$$

for the actual composition of the half-cells has a numerical value less than $K$, the equilibrium constant for the cell reaction.

EXERCISE 6.4:   Why? What happens to the value of $J$ when reaction 6-18 proceeds to the right?

Remember that $J$ need not equal $K$ since the reactants (in this cell $H_2$ and $Cu^{++}$) are kept separate, so that the over-all cell reaction need not be at equilibrium even though the two half-reactions are individually at equilibrium at the two electrode surfaces. In fact, since the chemical energy of the cell depends on Eq. 6-18 not being at equilibrium, the cell potential $E$ depends on the relative values of $J$ and $K$. The quantitative relationship between these quantities is the *Nernst Equation*,

$$E = \frac{RT}{nF} \ln \left( \frac{K}{J} \right) \tag{6-20}$$

where $R$ is the thermodynamic gas constant (8.314 joule/degree), $T$ is the absolute temperature of the cell, $F$ is Faraday's constant, the number of coulombs of charge on a mole of electrons (96,488 coulombs/mole), and $n$ is the number of moles of electrons appearing in the balanced half-reactions whose sum is the cell reaction. For the hydrogen-copper cell reaction as written in Eqs. 6-16–18, $n$ is 2. If we substitute numerical values for $R$, $T$, and $F$, and at the same time insert the constant 2.303 which converts natural to base-ten logarithms,[3] the Nernst Equation becomes

$$E = \frac{0.0592}{n} \log \left( \frac{K}{J} \right)$$

at $T = 298.2°K$ (25°C).

The Nernst Equation arises from thermodynamic reasoning similar to

[2]The proper quotient of activities was introduced in Chapter 1.
[3]See Appendix 1.

that which produces equilibrium constants, and its derivation is beyond the scope of this book. However, we can make it look less arbitrary. The product $RT$ is a kind of natural unit of energy per mole of chemical change. The electrical energy of the cell, on the other hand, is measured by the amount of electrical work it can do. For every mole of $Cu^{++}$ reduced in the cell, two Faradays of electricity (two moles of electrons) pass through the electrodes and can be made to pass through an external circuit against an opposing voltage less than or equal to the cell's voltage. This external voltage could be the voltage drop across a resistor such as a light bulb, or the resistive voltage drop and "back emf" of an electric motor, or the like. The details of the external circuit do not concern us. Electrical work is measured as the product of transported charge and the voltage through which it is transported. For our cell, two Faradays of charge may be transported through an opposing voltage as large as that of the cell, 0.34 V; and for any other cell, $n$ Faradays of charge may be transported through an opposing voltage as large as $E$, the voltage of the cell. The electrical energy available from any cell, then, is the product $nFE$. Rearranging Eq. 6-20, we find that the electrical energy of the cell in multiples of the natural unit $RT$ is given by

$$\frac{nFE}{RT} = \ln\left(\frac{K}{J}\right) \qquad (6\text{-}21)$$

which emphasizes the close relationship between the electrical and chemical energies of the cell.

EXERCISE 6.5: What value of $E$, and what electrical energy, may be produced from the cell for the case of $J = K$? Explain this result.

**Standard potentials:** As another mental experiment on the Nernst Equation, suppose that we choose values of the activities of products and reactants in the cell reaction such that $J = 1.000 \ldots$ . One way (though not the only way) to do this would be to prepare the half-cells so that each activity in the proper quotient is unity. Unit activity is defined by physical chemists as a "standard state," and any physical quantity associated with a reaction in which all products and reactants are in their standard states is called the standard value of that quantity. The standard-state cell potential is the *standard potential*, $E^0$. From the condition $J = 1$, the Nernst Equation produces the neat relationship:

$$E^0 = \frac{RT}{nF} \ln K = \frac{0.0592}{n} \log K \qquad (6\text{-}22)$$

Equation 6-22 provides a useful connection between the new world of cell potentials and the familiar one of equilibrium constants.

One further algebraic move brings us to the most commonly encountered form of the Nernst Equation. If we rewrite Eq. 6-20 by breaking the logarithm into two parts, we find

$$E = \frac{RT}{nF}(\ln K - \ln J)$$

$$= \frac{RT}{nF}\ln K - \frac{RT}{nF}\ln J$$

But in view of Eq. 6-22, this becomes

$$E = E^0 - \frac{RT}{nF}\ln J \tag{6-23}$$

Equation 6-23 has the virtue of including only electrochemical constants, and of relating the cell potential $E$ directly to the standard potential $E^0$. For this reason, we will generally use the Nernst Equation in the form of Eq. 6-23 when the discourse is more practical than philosophical.

EXERCISE 6.6:   Write the Nernst Equation for the potential of the hydrogen-copper cell, and for a cell whose reaction is

$$MnO_2 + 2Fe(CN)_6^{4-} + 4H^+ = Mn^{2+} + 2Fe(CN)_6^{3-} + 2H_2O$$

supplying the proper quotient of activities in each case.

**The Nernst Equation and relative potentials of half-cells:**   The Nernst Equation clearly indicates that the potential of a particular cell depends on the composition of both half-cell solutions. Yet it is often of interest, and always easier, to focus attention on only one of the half cells while the composition and potential of the other are presumed to be held constant. In this case, the latter half-cell is serving as a *reference electrode*, and the Nernst Equation may be simplified as follows.

The universally chosen reference is the hydrogen electrode, in which the equilibrium

$$2H^+ + 2e^- = H_2(g) \tag{6-24}$$

is established at unit activity for $H^+$ and $H_2(g)$. This half-cell is in its standard state, so that its potential is its standard potential. The potential of this electrode will share the fate of many arbitrary numerical references in having a defined value of zero.

EXERCISE 6.7:   What is the Celsius temperature of melting ice? The longitude of Greenwich, England? The altitude of sea level?

That is, for the standard-state hydrogen electrode (the Standard Hydrogen Electrode, or S.H.E.) $E^0$ is precisely zero volts.

Now, consider a cell of which one electrode, say number 1, is the S.H.E., and the other (number 2) a general half-cell in which the equilibrium

$$aA + ne^- = bB + cC \qquad (6\text{-}25)$$

is established. If the cell reaction is

$$aA + \frac{n}{2}H_2 = bB + cC + nH^+$$

the Nernst Equation for the cell is

$$E_{cell} = E_{cell}^0 - \frac{RT}{nF} \ln \frac{(B)^b(C)^c(H^+)^n}{(A)^a(H_2)^{n/2}}$$

Both $E_{cell}$ and $E_{cell}^0$ are differences in the potentials of electrodes 1 and 2 and can be broken into two parts with the following results:

$$E_2 - E_1 = E_2^0 - E_1^0 - \frac{RT}{nF} \ln \frac{(B)^b(C)^c}{(A)^a} - \frac{RT}{nF} \ln \frac{(H^+)^n}{(H_2)^{n/2}}$$

Since the hydrogen electrode is in its standard state, all of the terms in the last equation which refer to it (e.g., $E_1$, $E_1^0$, and the second ln term) are identically zero, either by convention or because the log of unity is zero. We may then rewrite the Nernst Equation, leaving out the terms that are zero, to obtain

$$E_2 = E_2^0 - \frac{RT}{nF} \ln \frac{(B)^b(C)^c}{(A)^a} \qquad (6\text{-}26)$$

Equation 6-26 says that we may write a Nernst-like equation for a single electrode from its half-reaction (Eq. 6-25) as long as we remember that all potentials contained in that equation are values relative to the corresponding potentials of the reference electrode. Thus, $E_2$ is not the absolute potential of electrode 2, but the potential of a cell of which one half-cell is the S.H.E. Likewise, $E_2^0$ is a standard potential on a scale in which the S.H.E. is chosen as the reference electrode.

This is a good point at which to return to our qualitative considerations of the effect of the activities of substances in a half-cell on the potential of the electrode in that half-cell. For the copper electrode the half-reaction

$$Cu^{++} + 2e^- = Cu$$

is at equilibrium at the copper surface. We had concluded that an increase in $(Cu^{++})$ should give the copper a more positive charge. As one last check on the plausibility of the Nernst Equation, we may ask whether it predicts the same effect. From the half-reaction, we may write, for the copper electrode,

$$E_{Cu} = E_{Cu}^0 - \frac{0.059}{2} \log \frac{1}{(Cu^{++})}$$

or

$$E_{Cu} = E_{Cu}^0 + \frac{0.059}{2} \log (Cu^{++}) \tag{6-27}$$

In either form, but more obviously in the second, Eq. 6-27 predicts that an increase in $(Cu^{++})$ should cause $E_{Cu}$ to become more positive.

You should be wondering what would have happened if the copper half-reaction had been written as an oxidation. An answer to that question will be forthcoming when we consider the conceptual and mathematical differences between physical, electrostatic potentials and thermodynamic potentials. Suffice it for now to assert that a prediction of the electrostatic potential will be correct when the Nernst Equation is derived as we have from the half-reaction written as a reduction.

**Other reference electrodes:**  The hydrogen half-cell has certain disadvantages for routine use as a reference electrode. It must be provided with a supply of $H_2$ at 1 atm pressure, which is not always convenient. Moreover, the rate at which the $H^+ - H_2$ equilibrium is established at the platinum surface is not great and is sensitive to the chemical state of that surface. The measured potential is thus sensitive to impurities in the half-cell solution or the $H_2$ gas and to the unavoidable passage of small currents through the electrode in the process of a potential measurement. These practical considerations have led to the use of a variety of alternative reference electrodes whose potentials have been carefully measured relative to the hydrogen electrode. The most common by far is a mercury-mercurous chloride half-cell containing a saturated solution of potassium chloride. The equilibria are

$$Hg_2Cl_2(c) + 2e^- = Hg(l) + 2Cl^-$$
$$+$$
$$2K^+$$
$$\parallel$$
$$2KCl(c) \tag{6-28}$$

Since all equilibria in 6-28 are tied to pure phases with constant activity, the positions of all equilibria are invariant except for the effect of temperature on the various equilibrium constants. This half-cell, containing a saturated solution of KCl, is called a *saturated calomel electrode,* or S.C.E. (calomel is an archaic name for $Hg_2Cl_2$). The potential of a S.C.E. is $+0.2415$ V *vs.* S.H.E. at 25°C. Concentrations of KCl other than a saturated solution may be used, establishing different equilibrium ($Hg_2^{++}$) with correspondingly different potentials. For example, the same half-cell containing 1 $F$ KCl has a potential of $+0.280$ V *vs.* S.H.E. at 25°C. If for any reason chloride must be excluded from the cell (for example, a reaction involving ions forming stable chloride complexes), a mercury-mercurous sulfate reference may be used.

## Formal Potentials

We have earlier noted that information about molar concentrations of ionic species may be more accessible experimentally, and in some situations more meaningful, than information about activities. We need a function analogous to the equilibrium quotient $Q$ which will relate cell potentials to the formal concentrations of solutes in the half-cells. Taking a Nernst Equation for a half-cell potential (Eq. 6-26) as our model, we will define the *formal potential* of an electrode as the measured potential (relative, as always, to a specified reference electrode) of that half-cell when the proper quotient of formal concentrations of reactants and products of the half-cell reaction is unity. For example, for the copper electrode the formal potential $E^{0'}$ is the measured potential of a copper electrode in equilibrium with 1 $F$ Cu($NO_3$)$_2$, or other salt of $Cu^{++}$. For the half-reaction

$$VO_2^+ + 2H^+ + e^- = VO^{++} + H_2O \qquad (6\text{-}29)$$

the formal potential is the measured potential of a platinum or other inert electrode at whose surface Eq. 6-29 is at equilibrium when the proper quotient of concentrations

$$J' = \frac{C_{VO^{2+}}}{(C_{VO^{2+}})(C_{H^+}^2)} \qquad (6\text{-}30)$$

has a value of exactly unity.

The relation of formal potentials to standard potentials is most easily seen by considering a simple metal-metal ion electrode, though the principles are the same for any type of electrode. For a copper electrode,

$$E_{Cu} = E_{Cu}^0 + \frac{RT}{2F} \ln (Cu^{++})$$

By the definition of the activity coefficient (Chapter 1 or Appendix 2),

$$(Cu^{++}) = [Cu^{++}]f_{Cu^{++}} \qquad (6\text{-}31)$$

where $f_{Cu^{++}}$ is the activity coefficient of the cupric ion. If we substitute Eq. 6-31 into the Nernst Equation and expand the natural log term, we find

$$E_{Cu} = E_{Cu}^0 + \frac{RT}{2F} \ln [Cu^{++}] + \frac{RT}{2F} \ln f_{Cu^{++}}$$

This equation, which is always true, is true in the particular case of 1 $M$ cupric ion. In that case, the potential $E_{Cu}$ is by definition the formal potential, and $[Cu^{++}] = 1$, so that the Nernst Equation becomes

$$E_{Cu}^{0'} = E_{Cu}^0 + \frac{RT}{2F} \ln f_{Cu^{++}} \qquad (6\text{-}32)$$

Now $f_{Cu^{++}}$ is in general a function of the composition of the solution in which the cupric ion finds itself. Thus, the formal potential, like the equilibrium quotient, is not truly a constant, but depends on solution conditions. Because of this inherent ambiguity, the composition of the solution (identity and concentration of the major constituents) *must* be given along with any numerical value of a formal potential. The value is meaningless without the accompanying data. For example, the formal potential of the half-cell

$$Ce^{4+}, Ce^{3+} \mid Pt$$

is $+1.70$ V in 1 $F$ HClO$_4$, but $+1.28$ V in 1 $F$ HCl. The formal and standard potentials of this couple in pure water are practically impossible to measure because both ions, but particularly Ce$^{4+}$, are extensively hydrolyzed at pH 7.

For a given composition (for example, 1 $F$ HClO$_4$, etc.) of the main components of the solution, $E^{0'}$ will not be sensitive to small changes in the concentrations of other solution components, and will play a role similar to that of the standard potential. In view of Eq. 6-32, the Nernst Equation for the copper electrode may now be written

$$E_{Cu} = E_{Cu}^{0'} + \frac{RT}{2F} \ln [Cu^{++}] \qquad (6\text{-}33)$$

## Electrode Potentials in the Presence of Acid-base and Solubility Equilibria

With the aid of the formal potential to relate electrode potentials to molar and formal concentrations, we are now in a position to relate this chapter to the five preceding. Electrode potentials are often observed in the presence of other equilibria which affect the concentrations of one or more species in the half-reaction. The resulting potentials are always related to the potential of the simple electrode through the Nernst Equation. It will be convenient to consider separately Brønsted acid-base, Lewis acid-base, and solubility equilibria.

**pH-sensitive electrodes:** Any electrode whose half-reaction involves hydrogen ions must, by the Nernst Equation, show a potential dependent on pH. As one of the hundreds of possible examples, consider the half-reaction

$$MnO_4^- + 8H^+ + 5e^- = Mn^{++} + 4H_2O \qquad (6\text{-}34)$$

For a half-cell in which this equilibrium is established,

$$E = E^{0\prime} - \frac{0.059}{5} \log \frac{[Mn^{++}]}{[MnO_4^-][H^+]^8}$$

$$= E^{0\prime} - \frac{0.059}{5} \log \frac{[Mn^{++}]}{[MnO_4^-]} + \frac{0.059}{5} \log [H^+]^8$$

$$= E^{0\prime} - \frac{0.059}{5} \log \frac{[Mn^{++}]}{[MnO_4^-]} - \frac{8}{5} 0.059 \, pH \qquad (6\text{-}35)$$

In the last step of this development, the definition of pH and the fact that $\log [H^+]^8 = 8 \log [H^+]$ were combined.

EXERCISE 6.8:   On a graph of the half-cell potential *vs*. pH, what would be the shape of Eq. 6-35? Such graphs are useful in mapping out the redox equilibria of a particular couple and will be considered further in the problems at the end of this chapter.

Because of the direct relationship between pH-dependent electrode potentials and half-cell pH, several have been used as convenient experimental measurements of pH. Pre-eminent among these is the hydrogen electrode.

$$H^+ \mid Pt, H_2$$

EXERCISE 6.9:   Write the Nernst Equation for this electrode in terms of the half-cell pH.

A half-cell containing a sparingly soluble one-to-one complex of quinone with hydroquinone ("quinhydrone") is also useful.

EXERCISE 6.10:   The half-reaction for a quinhydrone electrode is given in Eq. 6-9. Because they are added in the form of a one-to-one complex, the concentrations of quinoné and hydroquinone in solution are equal. Write the Nernst Equation for this half-cell, and compare it to that for the $H^+ - H_2$ half-cell.

Finally, metal-metal oxide electrodes occasionally find use in the determination of pH, especially in nonaqueous solvents. The most frequently used electrode of this type is $Sb_2O_3$-$Sb$.

> EXERCISE 6.11: Write the half-reaction and the Nernst Equation for this electrode, and compare it to those for the hydrogen and quinhydrone electrodes.

All these electrodes have been virtually supplanted in practical pH measurements by glass membrane electrodes which operate through an ion-exchange equilibrium, and which we will discuss briefly later in this chapter.

**Complex formation and metal ion electrodes:**    Suppose that a metal electrode is in equilibrium with metal ions, which are themselves in equilibrium with a series of complex ions:

$$M^{n+} + ne^- = M(c)$$
$$+$$
$$L$$
$$\parallel$$
$$ML = ML_2 = ML_3 = \text{etc.} \qquad (6\text{-}36)$$

Regardless of the presence of the ligand, the metal-metal ion equilibrium continues to be described by the Nernst Equation, howbeit at a small $[M^{n+}]$ if the complexes $ML$, $ML_2$, $ML_3$, etc. are stable:

$$E_M = E_M^{0\prime} + \frac{0.059}{n} \log [M^{n+}]$$

$$= E_M^{0\prime} + \frac{0.059}{n} \log C_M \alpha_M \qquad (6\text{-}37)$$

If the equilibrium ligand concentration and the stepwise formation quotients for the series of complexes are known, $\alpha_M$ may be calculated by the methods developed in Chapter 3.

If the necessary data to calculate $\alpha_M$ are not known, the definition of the formal potential is broad enough to include the unknown $\alpha$:

$$E_M = E_M^0 + \frac{0.059}{n} \log (M^{n+})$$

$$= E_M^0 + \frac{0.059}{n} \log C_M \alpha_M f_M$$

$$= E_M^{0\prime} + \frac{0.059}{n} \log C_M \qquad (6\text{-}37)$$

where

$$E_M^{0\prime} = E_M^0 + \frac{0.059}{n} \log \alpha_M f_M \qquad (6\text{-}38)$$

In this case, the formal potential includes the value of $\alpha_M$ which, although unknown, is operationally fixed when the composition of the solution, including the formal concentration of the ligand L, is specified.

EXERCISE 6.12: By what factors is the formal potential for the $Ce^{4+}$, $Ce^{3+}$ couple in 1 $F$ HCl related to the standard potential? How do you expect these factors to compare for the two ions?

**Potentials in the presence of solubility equilibria:** We had a brief look at metal-metal salt electrodes earlier. We are now in a position to relate their standard potentials to those of the corresponding metal-metal ion electrodes. Consider for example the electrode

$$SO_4^{--}, Hg_2SO_4(c) \mid Hg(l) \qquad (6\text{-}39)$$

As we observed, the concentration of mercurous ion in this half-cell is governed by the $Q_{sp}$ of $Hg_2SO_4$ and the sulfate ion concentration. Writing the half-reaction for the half-cell as it stands in 6-39, we would say

$$Hg_2SO_4(c) + 2e^- = Hg(l) + SO_4^{--} \qquad (6\text{-}40)$$

and the Nernst Equation would be

$$E_{Hg} = E_{Hg_2SO_4,Hg}^{0\prime} - \frac{0.059}{2} \log [SO_4^{--}] \qquad (6\text{-}41)$$

EXERCISE 6.13: Why is $[SO_4^{--}]$ the only factor in the log term? Should it be in the denominator? What effect would an increase in $[SO_4^{--}]$ have on the electrode potential according to Eq. 6-40? Does this agree with 6-41?

However, we also have faith that if we knew the $[Hg_2^{++}]$ in this half-cell, we could calculate $E_{Hg}$ through the equally valid Nernst Equation

$$E_{Hg} = E_{Hg_2^{++},Hg}^{0\prime} + \frac{0.059}{2} \log [Hg_2^{++}] \qquad (6\text{-}42)$$

Now, in the presence of $Hg_2SO_4(c)$ and $SO_4^{--}$, it is always true that

$$[Hg_2^{++}] = \frac{Q_{sp}}{[SO_4^{--}]}$$

or

$$E_{Hg} = E^{0'}_{Hg_2^{++},Hg} + \frac{0.059}{2} \log \frac{Q_{sp}}{[SO_4^{--}]}$$

$$E_{Hg} = E^{0'}_{Hg_2^{++},Hg} + \frac{0.059}{2} \log Q_{sp} - \frac{0.059}{2} \log [SO_4^{--}] \qquad (6\text{-}43)$$

Equation 6-43, like Eq. 6-41, provides a method for calculating the potential of the mercury-mercurous sulfate electrode from a knowledge of the sulfate concentration. Since they both apply to the same electrode, they must predict the same value of $E_{Hg}$. A hard look at the right-hand sides of these two equations reveals that the constant term on the right-hand side of Eq. 6-41 must equal the sum of the constant terms on the right-hand side of Eq. 6-43.

EXERCISE 6.14:    Why?

Then

$$E^{0'}_{Hg_2SO_4,Hg} = E^{0'}_{Hg^{2++},Hg} + \frac{0.059}{2} \log Q_{sp} \qquad (6\text{-}44)$$

which is a pleasingly simple relationship between these three fundamental quantities.

EXERCISE 6.15:    This derivation in terms of formal potentials and equilibrium quotients could be paralleled by one in terms of standard potentials and equilibrium constants. It would be good practice for you.

Naturally, the derivation could also have been done in terms of a general insoluble salt $M_xY_y$; it is applicable to any metal-metal salt electrode. Equation 6-44 provides a remarkably convenient experimental method of determining solubility products, since all that is required is the evaluation of two formal (or, if the activity product is required, standard) potentials.

Once the formal potential of the metal-metal salt electrode is known, Eq. 6-41 or its analog permits the use of such an electrode as an anion-specific electrode; that is, the potential of the electrode is a measure of the concentration of anion (in our example, $SO_4^{--}$) in an unknown solution. For example, a silver-silver chloride electrode has been used to determine the chloride concentrations of seawater and biological fluids.

EXERCISE 6.16:    Write the Nernst Equation for this electrode and show how its potential depends on $[Cl^-]$. Since $E^{0'}$ is a function of the solution composition, what measurements would have to be made to evaluate it?

As is the case with pH-sensitive electrodes, metal-metal salt electrodes have begun to be superseded by ion exchange membrane electrodes for analytical purposes.

## Thermodynamic Potentials

In our first discussion of the copper-hydrogen cell, which led to the presentation of the Nernst Equation, we found that we could predict the spontaneous direction of the cell reaction from the fact that the copper electrode is more positive than the hydrogen electrode in that cell. It would be good to have a scheme for presenting this information a bit more compactly than our extended discussion. The scheme which chemists use flows from the Nernst Equation as we first encountered it, Eq. 6-20,

$$E_{\text{cell}} = \frac{RT}{nF} \ln \left( \frac{K}{J} \right)$$

The spontaneous direction of the cell reaction may be concluded from the relative values of $K$ and $J$: if $J < K$, then the reaction will proceed from left to right, and if $J > K$, the spontaneous direction will be from right to left.

EXERCISE 6.17:   Write any chemical reaction and confirm these conclusions. (How must the reaction proceed so that $J$ may approach $K$?)

But, by the properties of logarithms, if $J < K$, $E_{\text{cell}}$ is positive, and if $J > K$, $E_{\text{cell}}$ is negative. If we had a convention for the *direction* of reaction to be associated with a particular schematic cell, and a convention for determining the sign of $E_{\text{cell}}$, we could use this sign as a *code* to indicate whether or not the direction associated with the schematic cell as written is the spontaneous one. These conventions, which constitute part of the thermodynamic description of electrochemical cells, are the following:

(1) With any schematic cell we will associate the cell reaction which results from oxidation at the left-hand electrode and reduction at the right-hand electrode.

(2) The *thermodynamic* potential of the schematic cell will always be the difference in the *electrostatic* potentials of the real electrodes, taken in the sense $E_{\text{(right-hand electrode of schematic cell)}} - E_{\text{(left-hand electrode)}}$. For example, in a copper-zinc cell the copper electrode is roughly one volt more positive (for reasonable concentrations of copper and zinc ions) than the zinc electrode. The schematic cell

$$\text{Zn} \mid \text{Zn}^{++} \parallel \text{Cu}^{++} \mid \text{Cu} \qquad (6\text{-}45)$$

thus has a thermodynamic potential of $+1$ V and corresponds to the cell reaction

$$Zn + Cu^{++} \to Zn^{++} + Cu \qquad (6\text{-}46)$$

which is the spontaneous direction for this reaction. (Since we are now discussing reactions which are *not* at equilibrium, but which will seek equilibrium in a certain direction, it is important to replace the = sign with an arrow.)

EXERCISE 6.18:   Does the direction given for Eq. 6-46 correspond to your chemical experience? If you are not sure, try to decide what would happen if the two electrodes of the real cell were connected with a wire. Drop a piece of zinc into a cupric sulfate solution and observe developments.

On the other hand, the schematic cell

$$Cu \mid Cu^{++} \parallel Zn^{++} \mid Zn$$

is associated with the *nonspontaneous* reaction

$$Cu + Zn^{++} \to Cu^{++} + Zn \qquad (6\text{-}47)$$

and has a potential of $-1$ V.

EXERCISE 6.19:   Is $J$ larger or smaller than $K$ for Eq. 6-46? For Eq. 6-47? Are your answers to these questions consistent (through the Nernst Equation) with the sign of $E_{cell}$?

It is important to note that the sign of $E$ has significance only for a schematic cell and its conventionally associated reaction. There is no particular convention for the sign of a real cell's potential, since there is no particular direction in which to subtract the potential of one electrode from that of another. For example, the voltage of a flashlight battery is generally given as 1.5 V with no meaning attached to the implicit positive sign. *The sign of the potential for the schematic cell and its associated reaction is a "yes" ($+$) or "no" ($-$) code answer to the question, "Does the reaction conventionally associated with this schematic cell proceed spontaneously to the right?"*

**The sign of $E^0$ and the value of $K$:**   The convention relating the sign of a cell potential and the spontaneous direction of a reaction may be extended to the informative case of $J = 1$; that is to the equation

$$E^0 = \frac{RT}{nF} \ln K$$

If $E^0$ for any schematic cell is positive, we may immediately conclude that for the associated cell reaction, $K > 1$; and vice versa.

**Thermodynamic standard potentials for half-reactions and schematic half-cells:** Just as the electrostatic potentials of single electrodes can be expressed on a conventional scale with a hydrogen reference electrode, so also can the thermodynamic potentials for half-cells and their associated half-reactions. For example, the cell

$$\text{Pt, } H_2 \mid H^+, SO_4^{--}, PbSO_4 \mid Pb$$

has a standard potential of $-0.356$ V. Such a standard potential is by convention $E^0_{\text{right}} - E^0_{\text{left}} = E^0_{\text{right}}$.
Thus we may associate with the *half-cell*

$$SO_4^{--}, PbSO_4 \mid Pb$$

and with its half reaction

$$PbSO_4 + 2e^- \rightarrow Pb + SO_4^{--} \qquad (6\text{-}48)$$

the *standard reduction potential* $-0.356$ V *vs.* S.H.E. Note that this is a standard reduction potential because it is the standard potential of a schematic cell which has the S.H.E. on the left (oxidation) and the lead-lead sulfate electrode on the right (reduction). Now for the schematic cell

$$\text{Pb} \mid PbSO_4, SO_4^{--}, H^+ \mid \text{Pt, } H_2$$

the $E^0$ is opposite in sign:

$$E^0_{\text{cell}} = +0.356 \text{ V}$$

corresponding to the cell reaction

$$Pb + 2H^+ + SO_4^{--} \rightarrow PbSO_4 + H_2$$

Since the half-reaction

$$2H^+ + 2e^- \rightarrow H_2$$

has a conventional $E^0$ of zero, we conclude that the half-reaction

$$Pb + SO_4^{--} \rightarrow PbSO_4 + 2e^-$$

has a standard *oxidation potential* of $+0.356$ V *vs.* S.H.E. Thus, the standard oxidation potential of a given half-cell is equal in magnitude and opposite in sign to its standard reduction potential in just the same way and for the same reason that the sign of a schematic cell's thermodynamic potential is equal in magnitude and opposite in sign to the thermodynamic potential of the reverse schematic cell.

Again, it is important to understand that an oxidation potential or a reduction potential is a thermodynamic potential, whose sign is a code answer to the question, "Does this half-reaction go spontaneously from left to right in a cell, the other half-cell of which is a hydrogen electrode?" Such a code sign is conceptually quite different from the conventional signs given to electrical charges. This must be true since the thermodynamic sign of a given half-reaction depends on the direction in which it is written, but the electrostatic sign of the electrode at which that half-reaction is at equilibrium is independent of how one twists it about in space.

Surprisingly enough, some people have found this double system of signs confusing. Not everyone is careful to distinguish between thermodynamic and electrostatic signs. European chemists tend to use electrostatic potentials, and American chemists thermodynamic ones, to convey the same information. In 1953, the International Union of Pure and Applied Chemistry (IUPAC) recommended that all potentials be given the sign corresponding to the electrostatic charge on the electrode, and that the associated half-reactions always be written as reductions. These recommendations result in an unambiguous sign for a particular electrode, since the thermodynamic code-sign for a reduction half-reaction *happens*, as we have noticed, to coincide with the electrostatic sign of the electrode. In view of the confusion which sometimes arises without some such agreement, and the mnemonic virtues of having only one sign for a given electrode, we will use and list potentials of reduction half-reactions. However, you would be mistaken to allow this practice to blur the profound conceptual distinction between the thermodynamic code-sign of the half-reaction and the electrostatic sign of the electrode. You will need to understand this distinction since many of the important publications (for example, Wendell Latimer's outstanding 1952 collection of electrochemical data *Oxidation Potentials*) use oxidation potentials, as do some current authors even 16 years after the recommendations of IUPAC.

## Ion Exchange Electrodes

We cannot leave a discussion of electrode potentials without a brief introduction to ion-exchange membrane electrodes, even though the mechanism by which they operate is quite different from that of the oxidation-reduction electrodes we have considered up to this point. Where the fundamental equilibrium of an oxidation-reduction electrode may be represented as

$$ox + ne^- = red$$

an ion-exchange electrode responds to the difference in the activities of the ion I in two solutions according to:

$$I_{(soln. 1)} = I_{membrane} \ldots I_{membrane} = I_{(soln. 2)} \tag{6-49}$$

The ion I establishes an ion-exchange equilibrium (Chapter 5) at the inner (solution 1) and outer (solution 2) surfaces of the ion-exchange membrane (Fig. 6-8). The result of these two equilibria is that an electrical voltage is established across the ion-exchange membrane. This voltage then becomes part of the total voltage difference observed between two reference electrodes in a cell of the form

reference 1 | solution 1 | membrane | solution 2 | reference 2    (6-50)

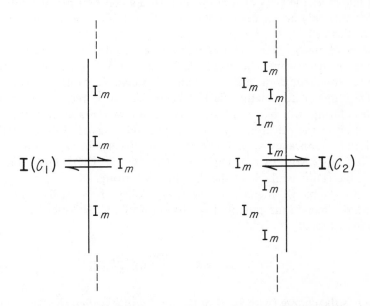

**Fig. 6-8.** Equilibrium at the two surfaces of an ion-exchange membrane. With $C_2$ greater than $C_1$, the ion I in the membrane ($I_m$) is at a higher activity at the right-hand surface than at the left, resulting in the establishment of an electrical potential across the membrane.

The voltage of the entire cell depends only on the voltage difference across the ion-exchange membrane, since the reference electrodes are so constructed as to be insensitive to the presence of ion I in the two solutions. Solution 1 is prepared with a constant activity of ion I, so that the voltage across the membrane depends only on the activity of I in solution 2, which may be a solution whose I content one wants to measure.

By far the most common application of this idea is the measurement of the pH of solutions (solution 2) in the cell

$$\text{Ag} \mid \text{AgCl, HCl} \left| \begin{array}{c} \text{glass} \\ \text{membrane} \end{array} \right| \text{solution 2} \mid \text{KCl(sat.)},\text{Hg}_2\text{Cl}_2 \mid \text{Hg} \qquad (6\text{-}51)$$

The Ag-AgCl electrode constitutes Reference 1 in the symbology of 6-50; the HCl solution is solution 1, whose hydrogen ion activity is constant; the glass membrane is an ion-exchange membrane nearly specific for hydrogen ions; the solution whose pH is to be measured is solution 2; and the saturated calomel electrode is reference 2. Since both references 1 and 2 in this cell are solutions of constant chloride ion activity, their potentials are invariant. The potential of the entire cell depends only on the potential across the glass membrane and, with a constant pH in solution 1, this depends only on the pH of solution 2. Figure 6-9 indicates the physical form of this cell.

There are many other such ion-exchange electrodes, each nearly specific for that ion with which the exchangeable sites in the ion-exchange membrane are supplied. Both cation- and anion-specific electrodes are available, and the scope of the method is just beginning to be realized after some decades during which the pH-sensitive glass membrane was the only application of the principle. Membrane electrodes, using glass, crystalline, and liquid membranes responding nearly specifically in each case to $H^+$, $Li^+$, $Na^+$, $Ag^+$, $Mg^{++}$, $Ca^{++}$, $Cu^{++}$, $F^-$, $Cl^-$, $Br^-$, $I^-$, $ClO_4^-$, $NO_3^-$, and $S^{--}$ are commercially available at the time of this writing.

All ion-exchange membrane electrodes obey a Nerst-like equation relating the membrane potential to the ratio of the activities of ion I in solutions 1 and 2:

$$E_{\text{membrane}} = k + \frac{0.059}{n} \log \frac{(\text{I})_1}{(\text{I})_2} \qquad (6\text{-}52)$$

where $n$ is the charge (with sign) of the ion I and $k$ (called the "asymmetry potential") is a small voltage which may depend on the recent history of the electrode, and so varies from day to day. The slight inconstancy of the asymmetry potential (for a typical glass pH electrode, $k$ may vary from day to day by an amount which corresponds to a few tenths of a pH unit) necessitates periodic standardization of the electrode with known solutions

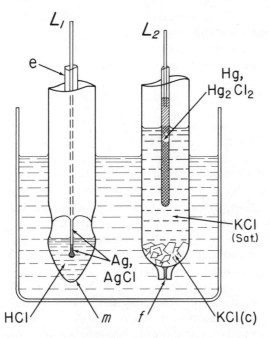

**Fig. 6-9.** A cell for determining the pH of the solution in the beaker. Compare cells 6.50 and 6.51. $m$ is the glass membrane across which the pH-dependent potential is established. The cell potential is measured across the leads $L_1$ and $L_2$ by means of an electronic voltmeter calibrated in pH (i.e., a pH meter). Because of the high resistance of the glass membrane, $L_1$ is electrically shielded by the conducting sleeve $e$ (which is isolated from $L_1$). The liquid junction between the KCl solution and the external solution whose pH is being measured is established at an asbestos fiber $f$. The construction of this cell is such that the two "electrodes" actually comprising, in the language of 6-50, reference 1, solution 1, membrane, and reference 2, may be transferred conveniently from one "solution 2" to another.

of I. A pH meter is standardized with buffers whose pH is known to within a small uncertainty.

The fact that there is a liquid junction between solution 2 (the unknown solution) and reference 2 in almost every case, and that this potential varies with the composition of solution 2, imposes an unavoidable uncertainty of the values of I activity measured in such cells as 6-50. In favorable cases, this may be as little as 0.01 log (I) units; but for solutions whose composition is very different from that of the solution used to standardize the electrode, the error may approach an order of magnitude. For this reason it is good practice, for example, to calibrate a pH meter with a buffer whose pH is close to that of the unknown solution.

## Irreversible Electrodes

The discussion throughout this chapter has assumed that the half-reactions

$$ox + ne^- = red$$

established at the electrode surfaces in the cell come to equilibrium very rapidly, so that the cell yields a potential in accord with the Nernst Equation. For many electrodes the rate of approach to equilibrium is so slow that the slight electrical current drawn through the electrodes by the potential-measuring device throws the half-reactions greatly out of equilibrium. Such electrodes are said to behave irreversibly. In this case the cell potential is always less than that predicted by the Nernst Equation, and cannot be interpreted to give information about the equilibrium conditions in the half-cell solutions.

For that matter this book, is discussing what is true at equilibrium, says nothing about what may be true of systems not at equilibrium. The predictions of thermodynamics (which is to say, the description of systems at equilibrium) can only say what the final state of any collection of matter will be. For a wide variety of systems (including living beings) the approach to equilibrium is slow and rich in interest.

## Suggestions for Further Reading

de Béthune, A., and N. A. S. Loud, *Standard Aqueous Electrode Potentials and Temperature Coefficients At 25°C.* Skokie, Illinois: Clifford A. Hampel, 1964. In addition to the very complete tables of data, there is a discussion of the applications of potential measurements to the deduction of other chemical information. Part of the discussion requires a knowledge of thermodynamics, but some will be of interest and use to readers of this book.

Freiser, H., and Q. Fernando, *Ionic Equilibria in Analytical Chemistry.* New York: John Wiley & Sons, Inc., 1963. Chapter XI is a good discussion of redox equilibria with copious use of graphs, including the pH-potential graphs introduced in problem 6.10. Chapter XVI applies redox equilibria to an explication of titrations involving oxidizing and reducing agents.

Kolthoff, I. M., P. J. Elving, and E. B. Sandell, eds., *Treatise On Analytical Chemistry.* New York: Interscience Encyclopedia, 1959. Chapters 9 (by Roger G. Bates) and 16 (by Frederick R. Duke) of Part I, Volume I discuss electrode potentials and redox equilibria at the exhaustive depth characteristic of this book. Chapter 15 (by Frederick R. Duke) is a treatment of the mechanisms of oxidation-reduction reactions which, since it describes the journey toward equilibrium, rather than conditions at equilibrium, would be a worthwhile complement to our discussion. An elementary treatment of chemical kinetics should be read before Duke's exposition is tackled.

Latimer, W., *The Oxidation States of the Elements and Their Potentials In Aqueous Solutions*, 2nd ed. Englewood Cliffs, N.J.: Prentice-Hall, 1952. Probably the most complete and helpful discussions of the uses and interrelationships of electrochemical data, as well as a very complete assembly of electrochemical data, element by element. Read at least parts of this book to see the kinds of conclusions that can be drawn from a relatively small number of observations.

Licht, T. S., and A. J. de Béthune, "Recent Developments Concerning The Sign of Electrode Potentials." *Journal of Chemical Education*, *34*, 433 (1957). Presents and discusses the IUPAC recommendations on sign conventions for potentials.

Lingane, J. J., *Electroanalytical Chemistry*, 2nd ed. New York: Interscience Publishers, Inc., 1958. Chapters I through VIII present and use in an analytical context the theory, practice, and lore of measuring and interpreting cell potentials. In particular, Chapter II, on common electrical measurements, should be useful to those whose educations have not included this art. Of all the works cited here, you will probably find Lingane's book the most artfully written and the most educational.

Rechnitz, G. A., "Ion-Selective Electrodes," *Chemical and Engineering News*, June 12, 1967. A critical comparison of the theory and behavior of ion-exchange membrane electrodes of all types. The exposition, which includes descriptions of commercial electrodes, is at a level accessible to the reader of this book.

# Problems

6.1.  For the following schematic cells, write the cell reaction and, from the data in Appendix 3, deduce $E^0$ for the cell and calculate the equilibrium constant for the cell reaction. For which schematic cells is the conventionally associated direction of reaction the spontaneous one?

A. $Pt \mid Fe^{3+}, Fe^{++} \parallel Ag^+ \mid Ag$

B. $Li \mid Li^+, F^- \mid Pt, F_2$

C. $Hg \mid HgO, OH^-, Zn(OH)_2 \mid Zn$

D. $Cd \mid Cd(OH)_2, OH^- \parallel H^+ \mid Pt, H_2$

E. $Au \mid quinone, hydroquinone \parallel I^-, I_2 (c) \mid Pt$

F. $Pb \mid PbSO_4(c), H_2SO_4 (aq), PbO_2 \mid Pb$

6.2.  For the following reactions (which are spontaneous in the direction indicated) write a schematic cell and identify the positive electrode.

$Cd + Pb^{++} \rightarrow Cd^{++} + Pb$

$Sn^{4+} + 2Ag \leftarrow Sn^{++} + 2Ag^+$

$Fe^{3+} + Ag + Br^- \rightarrow Fe^{++} + AgBr$

$Ag^+ + Br^- \rightarrow AgBr$

6.3.  A. Why does the potential of a cell approach zero as the cell reaction proceeds?

B. The cell in Fig. 6-1 (in different physical form) is often used in electronic instruments because its potential remains essentially constant during discharge. Write the schematic cell diagram, the cell reaction, and the equilibrium constant for this cell. Explain why the potential remains constant during discharge of the cell. Why can this cell not discharge forever at the same potential? Sketch a graph showing its cell potential as a function of time near the end of its life.

6.4. Write the cell reaction and calculate the equilibrium constant for the cell

$$Ag \mid Ag^+ \parallel Cl^-, AgCl \mid Ag$$

Compare the equilibrium constant to the tabulated value of $K_{sp}$ for AgCl in Appendix 3.

6.5. Cells of the type

$$Cd \mid Cd^{++}(C_1) \mid Cd^{++}(C_2) \mid Cd$$

are called concentration cells. If $C_2 > C_1$,
A. Which electrode is positive?
B. Write the reaction for this cell. Is this the spontaneous direction for this reaction? What is the value of $E^0$ for this cell?
C. Devise a concentration cell in which the reaction is

$$SO_4^{--}(C_2) \rightarrow SO_4^{--}(C_1)$$

6.6. What is the relationship between the cell of problem 6.4 and a silver ion concentration cell?

6.7. From the formation quotients for $HgCl^+$ and $HgCl_2(aq)$, calculate the $[Hg^{++}]$ in 1 $F$ $HgCl_2$ and from this information plus the standard potential of the $Hg^{++}$-$Hg$ electrode, estimate the standard potential of the electrode

$$HgCl_2(aq) \mid Hg$$

(The factor $(0.059/2) \log f_{Hg^{++}}$ will be very small.)

6.8. Why do the potentials of all metal-metal oxide electrodes show a pH dependence of $-0.0592$ V per pH unit regardless of the oxidation state of the metal in the oxide?

6.9. Reilley and Schmid [*Analytical Chemistry*, 30, 947 (1958)] have described a "pM" electrode for which the equilibria are

$$Hg^{++} + 2e^- = Hg(1)$$
$$+$$
$$Y$$
$$\parallel$$
$$HgY + M^{n+} = MY + Hg^{++}$$

where Y is a chelon, and $M^{n+}$ a metal ion whose chelate with Y is less stable than that of $Hg^{++}$. Beginning with the Nernst Equation for the $Hg^{++}$-Hg electrode, show that

$$E_{Hg} = E^{0\prime}_{Hg^{++},Hg} + \frac{0.059}{2} \log \frac{[HgY]Q_{MY}[M^{n+}]}{Q_{HgY}[MY]}$$

where $Q_{HgY}$ and $Q_{MY}$ are the formation quotients of HgY and MY. Observations of the potential of this electrode are used to follow the progress of titrations of $M^{n+}$ with Y (since the ratio $[M^{n+}]/[MY]$ changes greatly at the equivalence point of a titration) and to measure formation quotients of metal chelates in solutions containing known ratios of $M^{n+}$ to MY.

6.10. For copper, the following data apply:

$Cu^{++} + 2e^- \rightarrow Cu;\ E^0 = +0.337\ V$  (I)

$Cu(OH)_2 + 2e^- \rightarrow Cu + 2OH^-;\ E^0 = -0.219\ V$  (II)

A. Calculate the $K_{sp}$ of $Cu(OH)_2$.

B. On the same set of $E - pH$ axes, draw graphs of $E$ for half-reactions (I) and (II) as a function of pH. In (I), assume $(Cu^{++}) = 1$.

C. To the low pH side of the intersection of the $E$-pH lines for these two half-reactions, $E_{Cu(OH)_2,Cu}$ is more positive than $E_{Cu^{++},Cu}$. In this region of pH, what would be the *spontaneous* reaction in the cell

$$Cu \mid Cu(OH)_2, OH^- \parallel Cu^{++} \mid Cu?$$

D. What would be the spontaneous reaction in the above cell in the pH region above the intersection?

E. At what pH is the reaction

$$Cu(OH)_2 = Cu^{++} + OH^-$$

at equilibrium for $(Cu^{++}) = 1$?

F. Draw a vertical line on your graph at the pH you calculate for part E. Label areas on the graph where $Cu^{++}$, $Cu(OH)_2$, and Cu are the stable forms of copper, assuming unit activity for each.

# Some Mathematical Tools

Logarithms: A logarithm is an exponent. It is the exponent to which another number, called the *base*, must be raised to obtain a third number, the *antilogarithm*. Thus if

$$N = B^L$$

then $L$ is said to be the logarithm of $N$ to the base $B$;

$$L = \log_B (N)$$

Although $B$ may be any number, only two are commonly encountered: 10 and the natural constant $e$. If no base is indicated, 10 is generally intended; thus, since

$$100 = 10^2$$
$$\log_{10} (100) = 2$$

or, without the specifying subscript,

$$\log 100 = 2$$

Logarithms to the base $e$ are called natural logarithms, and are indicated by the notation ln. Thus if

$$a = e^x$$

then

$$x = \ln a$$

The relation between logarithms to the base 10 and those to the base $e$ is a simple one. Suppose that

$$\ln a = x$$

and

$$\log a = y$$

Then

$$a = 10^y = e^x$$
$$10 = e^{x/y}$$
$$\ln 10 = x/y$$

and

$$x = y \ln 10$$

That is,

$$\ln a = \log a \cdot \ln 10$$

The value of ln 10 is 2.303, so that

$$\ln a = 2.303 \log a, \text{ for any } a$$

**Logarithms of products and ratios:**    Suppose that

$$\log a = x$$

and

$$\log b = y$$

Then

$$a = 10^x$$
$$b = 10^y$$
$$a \cdot b = 10^x \cdot 10^y = 10^{(x+y)}$$

so that

$$\log(ab) = x + y = \log a + \log b$$

Now,

$$1/a = 1/10^x = 10^{-x}$$

so

$$\log 1/a = -x = -\log a$$

This result allows us to contemplate the logarithm of a ratio:

$$\log\left(\frac{a}{b}\right) = \log\left(a \cdot \frac{1}{b}\right)$$

$$= \log a + \log\left(\frac{1}{b}\right)$$

$$= \log a - \log b$$

**Logarithms of small numbers:**    The relationship between a number less than one and its logarithm is often confusing. Since $\log 1 = 0$, the log of any number less than 1 must be negative. Thus $\log \frac{1}{2} = -\log 2 = -0.301$, since $\log 2 = +0.301$. When the number is expressed in decimal or exponential notation, however, the situation appears different, though of course it cannot really be:

$$\log 0.5 = \log(5 \times 10^{-1})$$

$$= \log 5 + \log(10^{-1})$$

Inserting values for the logarithms, we arrive at the same result as before:

$$\begin{array}{ll} \log 5 & = +0.699 \\ \log(10^{-1}) & = -1.000 \\ \hline \log 0.5 & = -0.301 \end{array}$$

**"p" notation:**    Because small numbers arise frequently in the study of ionic equilibria (values of trace concentrations, equilibrium constants, etc.), a notation has been developed which, though sometimes confusing to the beginner, is timesaving in the long run. The negative of the logarithm of a quantity is denoted by the prefix p attached to that quantity. For example, the autoprotolysis constant of water is

$$K_w = 1.0 \times 10^{-14} \text{ at } 25°C$$

Then

$$pK_w = 14.00 \text{ at } 25°C$$

As the reader may note, the second method of expressing this quantity is somewhat more compact than standard exponential notation. The use of the *negative* of the logarithm obviates the otherwise ubiquitous negative sign if small quantities were to be expressed by their ordinary logarithm. When the quantity of interest is a concentration, the brackets around the symbol are conventionally omitted. Thus, the negative logarithm of $[Cl^-]$ is denoted $pCl$, not $p[Cl^-]$. Occasionally, a subscript is used to distinguish the negative logarithm of a species' activity from that of its concentration: thus $p_aH$ is $-\log(H^+)$, and $p_cH$ is $-\log[H^+]$. In this text, we will use an unsubscripted p to indicate $-\log$ molarity unless another interpretation is specifically indicated.

The student would do well to practice a few calculations of the following types:

EXAMPLE A1.1:   The $[H^+]$ in a solution is calculated to be $3.5 \times 10^{-6}$ *M*. What is the pH?

*Answer:* $pH = -\log[H^+]$

$$= -(\log 3.5 + \log(10^{-6})).$$

$$\log 3.5 = +0.544$$

$$\log(10^{-6}) = -6.000$$

$$\overline{\log(3.5 \times 10^{-6}) = -5.456}$$

$$pH = +5.456. \quad \text{(Note: } Not \text{ 6.456!)}$$

EXAMPLE A1.2:   From Table A3.3, the $pK_{sp}$ of $BaCO_3$ is 8.29. What is the $K_{sp}$? *Answer:* From our experience with Example A1.1, we note that when a number is expressed in standard exponential notation, the pre-exponential factor is always 1 or larger, so that its logarithm is positive. If the exponent is negative, the logarithm of the number is obtained as a difference. We must then reverse this process when confronted with a purely negative logarithm like a $pK$ or a pH. Thus

$$\log K_{sp} = -8.29$$

$$= -9.00 + 0.71$$

$$K_{sp} = 10^{-9.00} \times 10^{0.71}$$

from a standard log table, we find that $10^{0.71} = 5.1$ (i.e., we find that log $5.1 = 0.71$, and we read the table backwards!) So

$$K_{sp} = 5.1 \times 10^{-9}.$$

**Graphs:**   You are doubtless familiar with the use of graphs to depict equations. For example, the graph of the equation

$$y = x$$

on $x - y$ coordinates is a straight line passing through the origin and sloping upwards through all points for which $y = x$:

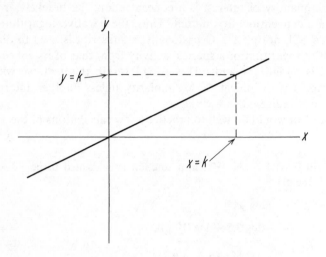

If

$$y = ax$$

the slope changes:

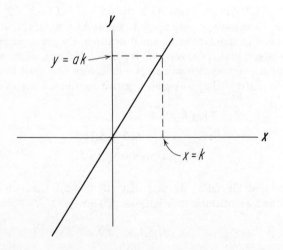

And if

$$y = ax + b$$

the line passes through $b$ at $x = 0$ (which is not necessarily the left-hand margin of the graph):

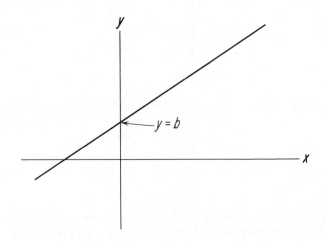

The purpose of these remarks is merely to remind you that an equation need not be obviously of the form $y = ax + b$ to graph as a straight line. For example, the equation

$$pH = pQ_a - \log \frac{C_A}{(C - C_A)}$$

is certainly not a linear-looking equation; on pH, $C_A$ axes, it is a curve:

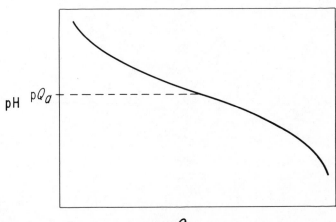

But on pH, $\log(C_A/(C - C_A))$ axes, it is a straight line with slope $-1$ and intercept $pQ_a$:

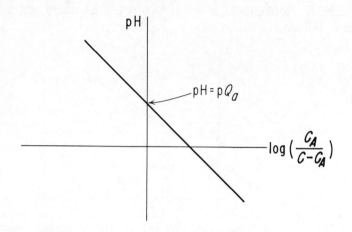

Straight lines have two virtues: they are easy to draw accurately and read for the information they contain (as for example, the logarithmic distribution graphs of Chapter 2); and experimental deviations from a straight-line graph are easier to detect than those from any other form of graph. We will often go out of our way to place equations in a form for graphing as a straight line by choosing proper coordinates [for example, Figures 4-1(a) and (b)].

# Some Thermodynamic Background Information

**A summary of the thermodynamic basis for equilibrium constants:** The following is not intended to teach you thermodynamics — a process which resists the most earnest efforts at simplification and summarizing — but to review, in outline form, the logical connection between the Second Law of Thermodynamics and the equilibrium constant as we have introduced it in this book. The second half of this appendix will then indicate how a knowledge of equilibrium constants and the ionic content of the solution will allow an approximate calculation of equilibrium *quotients* in that solution. It is not necessary to understand the thermodynamic basis of equilibrium constants in order to appreciate their consequences. The present discussion is intended as a review for those readers who already have some acquaintance with thermodynamic ideas.

The essential content of the Second Law may be succinctly stated: *No process may occur which results in a decrease in the entropy of the universe.* (This is a generalized statement of the lengthy order-disorder argument of Chapter 1; entropy ($S$) is a function which increases with increasing microscopic disorder.) Processes which may occur will result either in an increase in entropy (spontaneous, irreversible, real changes) or no change in the entropy of the universe (equilibrium, reversible changes). That is, the criterion for a possible process is

$$\Delta S_{univ} \geq 0 \qquad (A2\text{-}1)$$

Now $\Delta S_{univ}$ may be divided into two parts: $\Delta S$ of the system under study, and $\Delta S$ of its surroundings:

$$\Delta S_{syst} + \Delta S_{surr} \geq 0 \qquad (A2\text{-}2)$$

The term $\Delta S_{surr}$ is unsatisfactory because the detailed properties of the surroundings are of no interest and may be unknown. In a *constant-temperature* process, the surroundings have only two important properties: a definite temperature and an apparently infinite heat capacity. For example, when a reaction is carried out in a thermostatted bath, the products and reactants remain at, or close to, the temperature of the bath regardless of the uptake or release of heat by the reaction. Under these circumstances, the surroundings absorb heat reversibly from the system, since the heat flows from the system to the surroundings across an infinitesimal temperature gradient. The entropy increase of the surroundings is given by

$$\Delta S_{surr} = \frac{Q_{reversible}\ (surr)}{T_{surr}}$$

$$= \frac{-Q_{syst}}{T_{syst}}$$

At *constant pressure*, and as long as the system is not capable of doing work other than pressure-volume work against the pressure of the surroundings,

$$-Q_{syst} = -\Delta H_{syst}$$

So

$$\Delta S_{surr} = \frac{-\Delta H_{syst}}{T_{syst}}$$

and

$$\Delta S_{univ} = \Delta S_{syst} - \frac{\Delta H_{syst}}{T_{syst}} \geq 0 \qquad (A2\text{-}3)$$

Note that $\Delta S_{univ}$ is expressible as changes in the properties of the system only. We may drop the "syst" subscript as being implicit in what follows. We now summarize the relevant system properties ($\Delta S$, $\Delta H$, and $T$) in a single symbol after some rearrangement:

$$T\Delta S_{univ} = T\Delta S - \Delta H \geq 0$$

$$-T\Delta S_{univ} = \Delta H - T\Delta S \leq 0$$

Since $T$ is constant, $T\Delta S$ is

$$T(S_{final} - S_{initial}) = TS_{final} - TS_{initial}$$
$$= \Delta(TS)$$

The requirement of the Second Law becomes

$$-T\Delta S_{univ} = \Delta H - \Delta(TS) \leq 0$$

or

$$-T\Delta S_{univ} = \Delta(H - TS) \leq 0 \qquad (A2\text{-}4)$$

The collection of system properties $H - TS$ is given the symbol $G$ (its importance was first derived by J. Willard Gibbs, and it is called the Gibbs free energy):

$$-T\Delta S_{univ} = \Delta G \leq 0 \qquad (A2\text{-}5)$$

We must recall that Eq. A2-5 applies only to processes occurring at constant temperature and pressure, and with no work other than pressure-volume work being done by the system. Constraints other than these produce other combinations of system properties which serve as bookkeeping functions for the entropy change of the universe; but since those applying to Eq. A2-5 are by far the most commonly encountered (any reaction carried out in a thermostatted, open vessel, and not harnessed to do electrical, magnetic, surface, etc., work satisfies them), $G$ is the most commonly useful bookkeeping function.

Equation A2-5 states that only those processes resulting in a decrease in $G$ (under the indicated restraints) are possible; and that for a reaction proceeding at equilibrium (for example, an infinitesimal quantity of chemical change in an equilibrium mixture, or a mole of change in a very large equilibrium system), $\Delta G = 0$.

At this point, it would be most convenient to choose a chemical reaction to use as a subject for further developments, whose validity of course is independent of the particular reaction chosen. Consider again Chapter 1 and the reaction

$$2HCOOH = (HCOOH)_2$$

which we will symbolize, for simplicity, as

$$2M = D$$

(M for monomer, D for dimer). When two moles of M disappear and one mole of D appears,

$$\Delta G = \left(\frac{\partial G}{\partial n_D}\right)_{T,p,n_M} - 2\left(\frac{\partial G}{\partial n_M}\right)_{T,p,n_D}$$

where the partial derivatives give the dependence of the Gibbs free energy of the solution on the number of moles of dimer ($n_D$) and of monomer ($n_M$), respectively. The partial derivative $(\partial G/\partial n_i)_{T,p,n_j}$ is called the *chemical potential* of the $i$th component, and given the symbol $\mu_i$. Thus,

$$\Delta G = \mu_D - 2\mu_M \qquad (A2\text{-}6)$$

Each of the $\mu_i$ may be written as

$$\mu_i = \mu_i^0 + RT \ln a_i \qquad (A2\text{-}7)$$

where $a_i$ is the *activity* of the $i$th component, and $\mu_i^0$ is the chemical potential of that component in its standard state. Depending on the choice of standard state, $a_i$ may be defined as approaching $M_i$ or $m_i$ as these approach zero (ideal dilute solution reference state, generally chosen for dilute solutes) or as approaching the mole fraction $X_i$ as $X_i$ approaches 1 (pure substance reference and standard state, generally chosen for the solvent in a solution).

With these elaborations, $\Delta G$ for the formic acid dimerization becomes

$$\Delta G = \mu_D^0 + RT \ln a_D - 2\mu_M^0 - 2RT \ln a_M$$
$$= (\mu_D^0 - 2\mu_M^0) + RT \ln a_D - RT \ln a_M^2$$
$$= \Delta G^0 + RT \ln \frac{a_D}{a_M^2} \qquad (A2\text{-}8)$$

In the last step, Eq. A2-6 was applied to the standard state to identify the quantity in parentheses, and the *proper quotient of activities* (defined on page 5) appears as a result of combining the two logarithmic terms.

As we have remarked, at equilibrium $\Delta G = 0$:

$$0 = \Delta G^0 + RT \ln \left(\frac{a_D}{a_M^2}\right)_{\text{equi}}$$

or

$$-\Delta G^0 = RT \ln \left(\frac{a_D}{a_M^2}\right)_{\text{equi}} \qquad (A2\text{-}9)$$

But each of the $\mu^0$'s in $\Delta G^0$ is the value of $(\partial G/\partial n_i)$ under completely specified conditions: a particular standard state of composition and pressure; and is thus a function only of temperature. Thus $\Delta G^0$ itself is also a function only of temperature and is independent of the composition of the solution in which the reaction is taking place. The same must then be true of the right-hand side of Eq. A2-9. If $R$ and $T$ are constants, so must the equilibrium value of the proper quotient of activities, $(a_D/a_M^2)_{equi}$, be a constant. It is the equilibrium constant $K$. Thus,

$$-\Delta G^0 = RT \ln K \qquad (A2\text{-}10)$$

The relationship of $K$ to the entropy change of the universe which was our starting-point is worth summarizing: from Eq. A2-5,

$$\Delta S_{univ} = -\frac{\Delta G}{T}$$

If we imagine the reaction to occur with all reactants and products in their standard states, we may write

$$\Delta S^0_{univ} = -\frac{\Delta G^0}{T}$$

$$= R \ln K \qquad (A2\text{-}11)$$

Since $\Delta S^0_{univ}$ must be positive for the reaction to be spontaneous in the standard state, $K$ must be greater than 1; a conclusion we reached in Chapter 1 on common-sense grounds. A rearrangement of Eq. A2-11 shows the significance of $K$ more explicitly:

$$K = \exp(\Delta S^0_{univ}/R) \qquad (A2\text{-}12)$$

**Equilibrium constants and equilibrium quotients: corrections for nonideal behavior.** It is unfortunately true that the functional dependence of the chemical potential $\mu_i$ on the *molarity* of species $i$ in a solution is not so simple as Eq. A2-7, especially for ionic species. We are faced with finding a relationship between $a_i$, the activity of species $i$, and its molarity, $M_i$, and no completely accurate function which is general and simple has ever been found. As is a common practice in thermodynamics, the difficulties are epitomized in a single symbol:

$$f_i = \frac{a_i}{M_i} \qquad (A2\text{-}13)$$

where $f_i$ is called the *activity coefficient* of $i$. Then

$$a_i = M_i f_i$$

and

$$\mu_i = \mu_i^0 + RT \ln M_i + RT \ln f_i \qquad (A2\text{-}14)$$

For our monomer-dimer example, the equilibrium constant

$$K = \frac{a_D}{a_M^2} = \frac{(D)}{(M)^2}$$

becomes

$$K = \frac{[D]f_D}{[M]^2 f_M^2} = Q \cdot \frac{f_D}{f_M^2} \qquad (A2\text{-}15)$$

If all of the $f_i$ are unity, then $Q = K$, and the system is said to be behaving ideally. As we remarked in Chapter 1, the activities and molarities of uncharged species are not greatly different in most solutions. But for ions, $f_i$ may be quite different from unity, and a product of $f_i$'s which are different from unity may cause $Q$ and $K$ to be different by an order of magnitude. More typically, the nonideality of species in the numerator and denominator of a $K$ will partially cancel out, and $Q$ and $K$ will differ by a factor between one and two.

By a theoretical argument based on electrostatics and statistical mechanics, Debye and Hückel (1923) derived an approximate equation for ionic activity coefficients:

$$\log f_i = -A z_i^2 (\tfrac{1}{2} \Sigma M_j z_j^2)^{1/2} \qquad (A2\text{-}16^*)$$

In this equation, $A$ is a constant which depends on temperature and solvent ($A = 0.51$ for water at $25°C$), $z_i$ is the charge on $i$, and the summation of molarity times charge squared is taken over *all* ions $j$ in the solution, including $i$. This summation occurs so frequently that it is convenient to give it the symbol $I$, and the name *ionic strength;* then

$$\log f_i = -A z_i^2 I^{1/2}$$

The Debye-Hückel equation is a limiting law, correct in the limit of zero ionic strength, which predicts that the activity coefficient of any ion depends only on that ion's charge and the ionic strength of the solution in which it finds itself; that the identity of the other ions in the solution is unimportant beyond the specification of their charge; and that $f_i$ will decrease uniformly as $I$ increases. All of these predictions are true at extremely low ionic strength. But at ordinary concentrations, the behavior of real electrolytes deviates from the relatively simple predictions of the Debye-Hückel Limiting Law. Figure A2-1 depicts the behavior of the activity coefficients

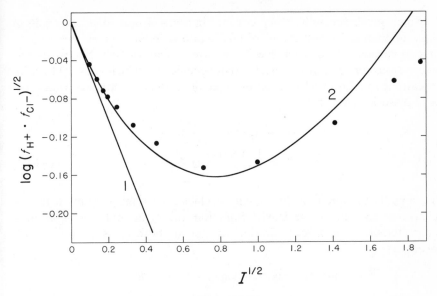

**Fig. A2-1**

of $H^+$ and $Cl^-$ in HCl solutions of various molalities (it is not possible to measure the activity coefficient of a single ion; from the definition of $\mu$, this would correspond to measuring the variation of the free energy of the solution as the concentration of a single ion is changed, an accomplishment proscribed by the electroneutrality principle).

Because the Debye-Hückel Limiting Law deviates widely from the observed behavior of electrolytes at moderate and high $I$, a variety of more or less empirical extensions of it have been proposed which fit the data with varying success at moderate $I$. It is always possible to fit any set of empirical data with an equation which contains a large number of adjustable empirical parameters; such equations are neither aesthetically appealing nor generally satisfactory, since they require the measurement and tabulation of otherwise meaningless constants for each ion studied. Because many situations confront the investigator with ions which have not been so studied, it would be best to have an equation which is at least approximately applicable to any ion. Of those which have been proposed, the most accurate and simple is the Davies Modification of the Debye-Hückel Limiting Law:

$$\log f_i = -Az_i^2\left[\frac{I^{1/2}}{1 + I^{1/2}} - 0.2I\right] \qquad (A2\text{-}16)$$

From Figure A2-1, it is clear that the fit of the Davies Equation to the data for HCl, while far from perfect, is more satisfactory either than the Debye-Hückel Limiting Law or than ignoring the difference between $M$ and $a$.

Since our discourse has been couched in terms of molarities and equilibrium quotients, the best means of taking into account nonideal behavior is not to calculate $f_i$ for each species involved in a reaction, but to derive expressions for the collections of activity coefficients which relate $Q$ to $K$ in equations like Eq. A2-15. As an example of an ionic reaction, consider the acid dissociation of HX:

$$HX + H_2O = X^- + H_3O^+$$

$$K_a = \frac{(H^+)(X^-)}{(HX)} = Q_a \cdot \frac{f_- \cdot f_+}{f_0} \qquad (A2\text{-}17)$$

To simplify the notation, the activity coefficients are labeled only with the charge on the ion; in the Davies Equation, this is the only distinguishing characteristic of a species. Taking the log of both sides of Eq. A2-17 produces

$$\log K_a = \log Q_a + \log f_+ + \log f_- - \log f_0$$

Inserting the Davies Equation for $\log f_i$, and collecting constant terms, we find

$$\log K_a = \log Q_a + [-A \cdot (+1)^2 - A \cdot (-1)^2 + 0]\left[\frac{I^{1/2}}{1 + I^{1/2}} - 0.2I\right]$$

Since the bracketed function of $I$ appears so frequently in calculations of this sort, let us give it a single symbol: $F_I$. Then we may rewrite the above equation

$$\log K_a = \log Q_a + (-A - A)F_I$$

or

$$\log Q_a = \log Ka + 2AF_I$$

in "p" notation, this becomes

$$pQ_a = pK_a - 2AF_I \qquad (A2\text{-}18)$$

which is not such a preposterous equation after all. In general, the relationship between $pQ_a$, $pK_a$, and $F_I$ will be of this simple sort. As another example, consider the complexation of $Fe^{3+}$ by $F^-$:

$$K_1 = \frac{(FeF^{++})}{(Fe^{3+})(F^-)} = Q_1 \cdot \frac{f_{2+}}{f_{+3} \cdot f_{-1}}$$

$$\log K_1 = \log Q_1 + \log f_{+2} - \log f_{+3} - \log f_{-1}$$

$$= \log Q_1 - A \cdot 4 \cdot F_I + A \cdot 9 \cdot F_I + A \cdot 1 \cdot F_I$$

So

$$\log Q_1 = \log K_1 - 6AF_I$$
$$= \log K_1 - 3.06F_I$$

EXERCISE A2.1:  Derive the relationship between $\log Q$, $\log K$, and $F_I$ for the equilibria

$$PbSO_4(c) = Pb^{++} + SO_2^{--}$$
$$HSO_4^- + H_2O = SO_4^{--} + H_3O^+$$

and

$$Cl_2 + 2I^- = I_2 + 2Cl^-$$

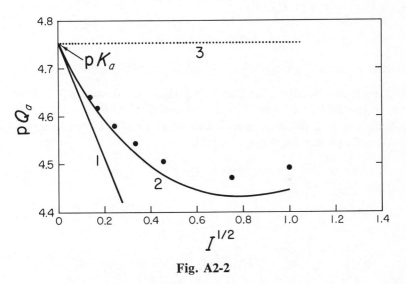

Fig. A2-2

In Figure A2-2, the prediction of the Davies Equation for $pQ_a$ of acetic acid is compared to experimental values over a wide range of ionic strength. Again, the fit of the Davies Equation to the data is not exact, but it would be preferable to no correction at all, in the absence of direct experimental measurements.

It might be of interest to examine the implications of all this for equilibrium calculations. For example, if I want to calculate the pH of a solution of acetic acid, how much does the result of that calculation depend on the value of $pQ_a$ that I use? Suppose, for simplicity, we use Equation 2-46* to calculate the pH of 0.1 $F$ acetic acid in the presence of 0.5 $F$ KCl ($I = 0.5$):

$$pH = \tfrac{1}{2}(pQ_a - \log C) \qquad (2\text{-}46^*)$$

(1) There is available in the chemical literature a series of careful measurements of $Q_a$ of acetic acid over a range of ionic strengths. The value for $I = 0.5$ is $pQ_a$ (exp) = 4.472. From Eq. 2-46*, pH = 2.74.

(2) If the experimental value of $pQ_a$ were not available, we would use the Davies Equation and the tabulated value of $pK_a$ (4.754) to calculate:

$$pQ_a(\text{Davies}) = pK_a - 2AF_I$$

$$= 4.754 - (2)(0.51)\left[\frac{0.5^{1/2}}{1 + 0.5^{1/2}} - (0.2)(0.5)\right]$$

$$pQ_a(\text{Davies}) = 4.432$$

$$pH = 2.72$$

(3) If we ignore the ionic-strength correction completely, we may approximate $pQ_a$ by $pK_a$:

$$pQ_a = pK_a = 4.754$$

$$pH = 2.88$$

To summarize, the use of the Davies Equation to calculate $pQ_a$ led to an error in the pH of 0.02 unit, which corresponds to a $+5\%$ error in [H$^+$]; ignoring the need for correction led to an error of 0.14 pH unit, which corresponds to an error in [H$^+$] of 38%.

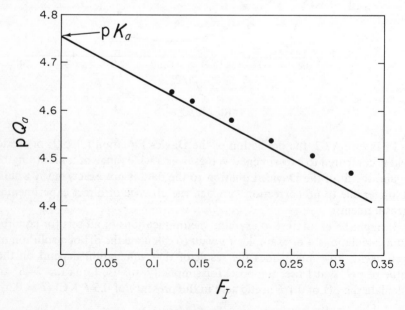

**Fig. A2-3**

EXERCISE A2.2:   How do these errors compare to that arising from the use of an approximate equation to calculate the pH?

Because the forces between ions which cause nonideal behavior are absent in the limit of infinite dilution (the ions are, usually, very far apart when the solution is dilute), the activity coefficient of any ion approaches unity as $I$ approaches zero. Consequently, $Q$ approaches $K$ as ionic strength decreases, and most $K$ values are obtained by extrapolating measured $Q$'s to zero ionic strength. Such an extrapolation is easier to make when the data can be plotted according to a linear equation. The Davies Equation predicts that $pQ$ should be a linear function of $F_I$, with an intercept of $pK$, and a slope that depends on the charges on the ions involved in the reaction. Experimental values of $pQ_a$ for acetic acid are plotted as a function of $F_I$ in Fig. A2-3. The line was drawn to have an intercept of 4.754, the accepted value of $pK_a$, and a slope of $-1.02$ ($= -2A$). Clearly, the experimental points approach agreement with the linear Davies plot quite closely as the ionic strength approaches zero.

# 3 Tables of Equilibrium Constants

The data in Tables A3.1, A3.2, and A3.3 were selected from *Stability Quotients*, a compilation by Lars G. Sillen and Arthur E. Martell, and constitute only a small fraction of the wealth of information obtainable from that source. All data refer to 25°C, unless otherwise noted, and are thermodynamic constants ($K$) if no value is given for the ionic strength $I$. If a value of $I$ is indicated, the data in that row are equilibrium quotients ($Q$) at the indicated ionic strength.

**TABLE 3.1.** DISSOCIATION CONSTANTS OF BRØNSTED ACIDS

| ACID | | $pK_1$ | $pK_2$ | $pK_3$ | $pK_4$ | $I$ |
|---|---|---|---|---|---|---|
| Acetic | | 4.7560 | | | | |
| $\alpha$-alanine (+1)[a] | | 2.340 | 9.870 | | | |
| $\beta$-alanine (+1) | | 3.55 | 10.23 | | | |
| Ammonium (+1) | | 9.245 | | | | |
| Anilinium (+1) | | 4.62 | | | | 0.05 |
| Arsenic | | 2.19 | 6.94 | 11.50 | | |
| Benzoic | | 4.01 | | | | 0.1 |
| *iso*-butyric | | 4.86 | | | | |
| *n*-butyric | | 4.8196 | | | | |
| Carbonic ($H_2CO_3$) | | 3.88 | 10.33 | | | |
| Carbonic ($CO_2 + H_2CO_3$) | | 6.35 | 10.33 | | | |
| Chloric | *ca.* | −2.7 | | | | |
| Chloroacetic | | 2.861 | | | | |
| Chromic | | — | 6.49 | | | |
| Citric | | 3.128 | 4.761 | 6.395 | | |
| 2,2'-diaminodiethylam- monium (+3) | | 3.64 | 8.74 | 9.80 | | |
| Diethanolammonium (+1) | | 9.00 | | | | 0.5 |

**TABLE 3.1.**   (*Continued*)

| ACID | | $pK_1$ | $pK_2$ | $pK_3$ | $pK_4$ | $I$ |
|---|---|---|---|---|---|---|
| Dimethylammonium (+1) | | 10.9 | | | | |
| Ethanolammonium (+1) | | 9.4980 | | | | |
| Ethylammonium (+1) | | 10.67 | | | | |
| Ethylenediamminetetra- | | | | | | |
| acetic acid (EDTA) | | 2.0 | 2.67 | 6.16 | 10.27 | 0.1 |
| Ethylenediammonium (+2) | | 7.18 | 9.96 | | | |
| Formic | | 3.7515 | | | | |
| Hydrazinium (+1) | | 7.94 | | | | |
| Hydrazoic | | 4.72 | | | | |
| Hydrocyanic | | 9.22 | | | | |
| Hydrofluoric | | 3.17 | | | | |
| Hydrogen peroxide | | 11.65 | | | | |
| Hydroselenic | | 3.89 | 11.0 | | | |
| Hydrosulfuric | | 6.99 | 12.89 | | | |
| Hydrotelluric | *ca.* | 2 | | | | |
| Hydroxylammonium (+1) | | 5.98 | | | | |
| Hypobromous | | 8.62 | | | | |
| Hypochlorous | | 7.53 | | | | |
| Hypoiodous | | 10.64 | | | | |
| Iodous | | 0.77 | | | | |
| Malonic | | 2.85 | 5.67 | | | |
| Mercaptoacetic | | 3.60 | 10.55 | | | |
| Nitric | *ca.* | −1.4 | | | | |
| Nitrilotriacetic (NTA) | | 1.66 | 2.95 | 10.28 | | |
| Nitrous | | 3.29 | | | | |
| Oxalic | | 1.25 | 4.285 | | | |
| Periodic | | 2.21 | | | | |
| Phenol | | 9.98 | | | | |
| Phosphoric | | 2.172 | 7.211 | 12.360 | | |
| Phthalic | | 3.14 | 5.40 | | | |
| Propionic | | 4.874 | | | | |
| Pyridinium (+1) | | 5.18 | | | | |
| Pyrophosphoric | | 1.52 | 2.36 | 6.60 | 9.24 | |
| Pyruvic | | 2.49 | | | | |
| Selenic | | — | 1.88 | | | |
| Sulfuric | | — | 1.96 | | | |
| Sulfurous | | 1.764 | 7.205 | | | |
| (±) Tartaric | | 3.04 | 4.37 | | | |
| Thiosulfuric | | 0.60 | 1.72 | | | |

[a]The electrical charge of the parent acid is indicated for all non-neutral species.

## TABLE 3.2. STABILITY CONSTANTS OF METAL-LIGAND COMPLEXES

*Note:* The order of ligands in this table is the same as that in Sillen and Martell: inorganic ligands, in the order OH⁻, then across the periodic chart from left to right; followed by organic ligands in order of increasingly complex empirical formulas. Given are the *logarithms* of the stepwise stability constants, or, if stepwise values are unavailable, of overall constants $\beta_n$.

| LIGAND | METAL | Log $K_1$ | Log $K_2$ | Log $K_3$ | Log $K_4$ | Log $K_5$ | Log $K_6$ | $I$ |
|---|---|---|---|---|---|---|---|---|
| OH⁻ | Li⁺ | 0.18 | | | | | | |
| | Mg²⁺ | 2.58 | | | | | | |
| | Ca²⁺ | 1.51 | | | | | | |
| | Fe²⁺ | 5.7 | | | | | | |
| | Fe³⁺ | 11.83 | | | | | | |
| | Cu²⁺ | 6.27 | 8.05 | | | | | |
| | Ag⁺ | 2.3 | 1.9 | | | | | |
| | Cd²⁺ | 4.9 | 2.2 | 0.69 | | | | |
| | Zn²⁺ | 4.36 | | | | | | |
| | Hg²⁺ | 10.2 | 11.6 | | | | | |
| CN⁻ | Fe²⁺ | | | | $\beta_4$ 30.3 | | $\beta_6$ 24 | |
| | Fe³⁺ | | | | 1.70 | | $\beta_6$ 31 | 3 |
| | Ni²⁺ | | $\beta_2$ 24.0 | | $\beta_4$ 16.76 | | | |
| | Cu¹⁺ | | | 4.59 | | | | |
| | Zn²⁺ | 5.48 | 5.12 | 4.63 | 3.55 | | | |
| SCN⁻ | Fe³⁺ | 3.03 | 1.94 | 1.4 | 0.8 | 0.02 | | |
| | Co²⁺ | 1.72 | | | | | | |
| | Ni²⁺ | 1.50 | | | | | | |
| | Cu²⁺ | 2.30 | 1.35 | 1.22 | 0.22 | | | |
| | Ag⁺ | 4.75 | 3.48 | 2.71 | 1.72 | | | |
| | Hg²⁺ | | $\beta_2$ 17.26 | | | | | |
| CO₃²⁻ | Ca²⁺ | 3.2 | | | | | | |
| NH₃ | Co²⁺ | 1.99 | 1.51 | 0.93 | 0.64 | 0.06 | ($t$ = 30°C) | |
| | Ni²⁺ | 2.80 | 2.05 | 1.66 | 1.3 | 1.2 | | |
| | Cu²⁺ | 4.01 | | | | −0.60 | | |
| | Cu²⁺ | 4.27 | 3.55 | 2.90 | 2.18 | −0.55 | | |
| | Ag⁺ | 3.37 | 3.84 | | | | | 0.1 |
| | Zn²⁺ | 2.18 | 2.25 | 2.31 | 1.96 | | ($t$ = 30°C) | 1.0 |

| Anion | Ion | | | | | | | |
|---|---|---|---|---|---|---|---|---|
| $NO_3^-$ | $Ag^+$ | −0.29 | | | | | | |
| | $Cd^{2+}$ | 0.31 | | | | | | |
| | $Pb^{2+}$ | 1.15 | | | | | | |
| $P_2O_7^{4-}$ | $Ca^{2+}$ | 5.60 | 2.3 | | | | | |
| | $Zn^{2+}$ | 8.7 | | | | | | |
| $S_2O_3^{2-}$ | $Ca^{2+}$ | 2.0 | 4.6 | | | | | |
| | $Ag^+$ | 8.9 | | | | | | |
| | $Zn^{2+}$ | 2.29 | 2.3 | | | | | |
| | $Cd^{2+}$ | 3.9 | | | | | | |
| $SO_4^{2-}$ ($t = 18°C$) | $Ca^{2+}$ | 2.28 | | | | | | |
| | $Ba^{2+}$ | 2.36 | | | | | | |
| | $Fe^{3+}$ | 4.15 | | | | | | |
| | $Pb^{2+}$ | 3.7 | | | | | | |
| | $Zn^{2+}$ | 2.3 | | | | | | |
| | $Cd^{2+}$ | 2.29 | | | | | | |
| $F^-$ | $Be^{2+}$ | 4.3 | 2 | | | | | |
| | $Mg^{2+}$ | 1.82 | | | | | | |
| | $Ce^{3+}$ | 4.00 | | | | | | |
| | $Th^{4+}$ | 8.65 | | | | | | |
| | $Fe^{3+}$ | 6.7 | 4.7 | | | | | |
| | $Al^{3+}$ | 7.00 | 6.05 | | | | | |
| | $Al^{3+}$ | 6.16 | 5.05 | 3.91 | 2.71 | 1.5 | | 0.53 |
| $Cl^-$ | $Fe^{3+}$ | 1.48 | 0.65 | −1.0 | | | | |
| | $Pd^{2+}$ | 6.1 | 4.6 | 2.4 | 2.6 | | | |
| | $Ag^+$ | 3.04 | 2.00 | 0.00 | 0.26 | −2.1 | −2.1 | |
| | $Cd^{2+}$ | 2 | | | | | | |
| | $Hg^{2+}$ | 6.74 | 6.48 | 0.85 | 1.00 | | | 0.5 |
| | $Tl^{1+}$ | 0.74 | | | | | | |
| | $Pb^{2+}$ | 1.62 | 0.82 | | | | | |
| $Br^-$ | $Fe^{3+}$ | 0.60 | | | | | | |
| | $Pd^{2+}$ | | | | $\beta_4$ 13.10 | | | |
| | $Pt^{2+}$ | | | | $\beta_4$ 20.5 | | | |
| | $Ag^+$ | 4.38 | 2.96 | 0.66 | 0.73 | | | |
| | $Hg^{2+}$ | 8.94 | 7.94 | 2.27 | 1.75 | | | 0.5 |

**TABLE 3.2.** (*Continued*)

| LIGAND | METAL | Log $K_1$ | Log $K_2$ | Log $K_3$ | Log $K_4$ | Log $K_5$ | Log $K_6$ | $I$ |
|---|---|---|---|---|---|---|---|---|
| I⁻ | $Fe^{3+}$ | 1.88 | 1.64 | 3.78 | 2.23 | | | 0.5 |
| | $Cd^{2+}$ | 2.28 | 10.95 | 0.77 | 0.55 | | | |
| | $Hg^{2+}$ | 12.87 | 1.15 | | | | | |
| | $Pb^{2+}$ | 2.0 | | | | | | |
| | $I_2$ | 2.89 | | | | | | |
| $IO_3^-$ | $Mg^{2+}$ | 0.72 | | | | | | |
| | $Ca^{2+}$ | 0.8 | | | | | | |
| | $Sr^{2+}$ | 0.96 | | | | | | |
| | $Ba^{2+}$ | 1.1 | | | | | | |
| | $Ag^+$ | 0.63 | 1.27 | | | | | |
| *Organic Ligands:* | | | | | | | | |
| $C_2O_4^{2-}$ (Oxalate) | $Ca^{2+}$ | 3.0 | | | | | | |
| | $Cd^{2+}$ | 4.00 | 1.77 | | | | | |
| | $Cu^{2+}$ | 6.19 | 4.04 | | | | | |
| | $Zn^{2+}$ | 5.00 | 2.36 | | | | | |
| $C_2H_3O_2^-$ (Acetate) | $Ag^+$ | 0.73 | −0.09 | | | | | |
| | $Ca^{2+}$ | 0.77 | | | | | | |
| | $Cu^{2+}$ | 2.24 | | | | | | |
| | $Cd^{2+}$ | 1.70 | | | | | | |
| $C_2H_8N_2$ (Ethylenediamine) | $Cd^{2+}$ | 5.41 | 4.5 | | | | | |
| | $Cu^{2+}$ | 10.5 | 9.1 | 4.3 | | | | |
| | $Ni^{2+}$ | 7.4 | 6.2 | | | | | |
| $C_2H_5O_2N$ (Glycine) | $Ag^+$ | 3.51 | 3.38 | | | | | |
| | $Cd^{2+}$ | 4.80 | 4.03 | | | | | |
| | $Co^{2+}$ | 5.23 | 4.02 | | | | | |
| | $Cu^{2+}$ | 8.62 | 6.97 | | | | | |
| | $Ni^{2+}$ | 6.18 | 4.96 | | | | | |
| | $Pb^{2+}$ | 5.17 | | | | | | |
| $C_3H_6O_2N$ (α-alanine) | $Cu^{2+}$ | 8.51 | 6.86 | | | | | |
| | $Ni^{2+}$ | 5.96 | 4.70 | | | | | |
| | $Zn^{2+}$ | 5.21 | 4.33 | | | | | |
| $C_4H_4O_2^-$ | $Cu^{2+}$ | 3.2 | 3.1 | 0.65 | 0.44 | | | 1.0 |

| Ligand | Cation | | | | $\mu$ |
|---|---|---|---|---|---|
| | $Zn^{2+}$ | 8.8 | 5.5 | | 0.1 |
| $C_4H_7O_2N_2^-$ (Dimethylglyoximate) | $Ni^{2+}$ | | $\beta_2$ 17.98 | | 0.5 |
| $C_4H_{11}O_2N$ (Diethanolamine) | $Ag^+$ | 2.69 | 2.79 | | |
| $C_5H_5N$ (Pyridine) | $Ag^+$ | 2.0 | 2.1 | | |
| $C_5H_8O_2$ (Acetylacetone) | $Fe^{3+}$ | 11.4 | 10.7 | 4.6 | |
| | $Ni^{2+}$ | 5.92 | 4.57 | | |
| $C_6H_6$ (Benzene) | $Ag^+$ | 0.38 | 0.67 | | |
| $C_6H_5O_7^{3-}$ (Citrate) | $Ca^{2+}$ | 4.90 | 3.12 | | |
| | $Cd^{2+}$ | 5.36 | | | |
| | $Cu^{2+}$ | 14.21 | | | |
| $C_6H_6O_6N^{3-}$ (Nitrilotriacetate, NTA) | $Ca^{2+}$ | 7.60 | | | 0.1 |
| | $Cu^{2+}$ | 13.10 | | | |
| | $Mg^{2+}$ | 6.5 | | | |
| $C_8H_4O_4^{2-}$ (Phthalate) | $Cu^{2+}$ | 3.46 | 1.37 | | |
| | $Ca^{2+}$ | 2.43 | | | |
| $C_8H_{23}N_5$ (Tetraethylenepentamine) | $Cd^{2+}$ | 14.0 | | | 0.1 |
| | $Cu^{2+}$ | 22.9 | | | 0.1 |
| | $Hg^{2+}$ | 27.7 | | | 0.1 |
| | $Mn^{2+}$ | 7.0 | | | 0.1 |
| | $Ni^{2+}$ | 17.8 | | | 0.1 |
| | $Zn^{2+}$ | 15.4 | | | |
| $C_{10}H_{12}O_8N_2^{4-}$ (Ethylenediaminetetraacetate, EDTA) | $Ba^{2+}$ | 7.7 | | | 0.1 |
| | $Mg^{2+}$ | 9.1 | | | 1.0 |
| | $Ca^{2+}$ | 16.4 | | | 0.065 |
| | $Fe^{3+}$ | 25.7 | | | 0.1 |
| | $Hg^{2+}$ | 22.15 | | | |
| | $Pb^{2+}$ | 17.9 | | | |

## TABLE 3.3.    SOLUBILITY PRODUCT CONSTANTS

*Note:* Salts are listed according to the anion, in the same order as these appear in Table 3.2.

| ANION | METAL | $pK_{sp}$ |
|---|---|---|
| $OH^-$ | $Be^{2+}$ | *ca.* 21 |
| | $Mg^{2+}$ | 10.6 |
| | $Ca^{2+}$ | 5.04 |
| | $Cr^{3+}$ | *ca.* 30 |
| | $Mn^{2+}$ | 12.80 |
| | $Fe^{2+}$ | 14.84 |
| | $Fe^{3+}$ | 39.4 |
| | $Co^{2+}$ | 14.9 |
| | $Cu^{2+}$ | 18.2 |
| | $Ag^+$ | 7.73 |
| | $Cd^{2+}$ | 14.4 |
| | $Zn^{2+}$ | 17 |
| | $Hg^{2+}$ | 25.4 |
| | $Al^{3+}$ | 32.4 |
| | $Pb^{2+}$ | *ca.* 15 |
| $CrO_4^{2-}$ | $Ba^{2+}$ | 9.93 |
| | $Cu^{2+}$ | 5.44 |
| | $Ag^+$ | 11.89 |
| | $Hg^{2+}$ | 8.70 |
| | $Pb^{2+}$ | 12.5 |
| $CN^-$ | $Cu^+$ | 19.49 |
| | $Ag^+$ | 15.92 |
| | $Zn^{2+}$ | 12.59 |
| $SCN^-$ | $Cu^+$ | 14.32 |
| | $Ag^+$ | 11.97 |
| | $Hg^{2+}$ | 19.52 |
| $CO_3^{2-}$ | $Mg^{2+}$ | 7.46 |
| | $Ca^{2+}$ | 8.54 |
| | $Sr^{2+}$ | 9.96 |
| | $Ba^{2+}$ | 8.29 |
| | $Fe^{2+}$ | 10.68 |
| | $Pb^{2+}$ | 10.78 |
| | $Cd^{2+}$ | 11.28 |
| $PO_4^{3-}$ | $Ca^{2+}$ | *ca.* 26 |
| | $Pb^{2+}$ | 42.10 |
| $SO_4^{2-}$ | $Ca^{2+}$ | 5.92 |
| | $Sr^{2+}$ | 6.49 |
| | $Ba^{2+}$ | 9.99 |
| | $Ra^{2+}$ | 14 |
| | $Ag^+$ | 4.80 |
| | $Hg_2^{2+}$ | 6.32 |
| | $Pb^{2+}$ | 7.79 |
| $F^-$ | $Ca^{2+}$ | 10.31 |
| | $Sr^{2+}$ | 8.61 |
| | $Pb^{2+}$ | 7.57 |
| $Cl^-$ | $Ag^+$ | 9.75 |
| | $Hg_2^{2+}$ | 17.88 |
| | $Tl^+$ | 3.75 |
| | $Pb^{2+}$ | 4.67 |
| $Br^-$ | $Ag^+$ | 12.34 |
| | $Hg_2^{2+}$ | 22.24 |
| | $Pb^{2+}$ | 4.41 |

**TABLE 3.3.**  (*Continued*)

| ANION | METAL | $pK_{sp}$ |
|---|---|---|
| $I^-$ | $Ag^+$ | 16.081 |
| | $Hg_2^{2+}$ | 28.31 |
| | $Pb^{2+}$ | 8.06 |
| $IO_3^-$ | $Ag^+$ | 7.52 |
| | $Zn^{2+}$ | 5.41 |
| | $Tl^+$ | 5.51 |
| $C_2O_4^{2-}$ | $Ca^{2+}$ | 8.64 |
| $C_4H_8O_2N$ (Dimethylglyoxime) | $Ni^{2+}$ | 23.66 |

**TABLE 3.4.**  SOME STANDARD AND FORMAL REDUCTION POTENTIALS[1]

| HALF REACTION | $E°$, VOLTS | FORMAL POTENTIAL, VOLTS |
|---|---|---|
| $F_2 + 2H^+ + 2e \rightleftharpoons 2HF$ | 3.06 | |
| $O_3 + 2H^+ + 2e \rightleftharpoons O_2 + H_2O$ | 2.07 | |
| $S_2O_8^{2-} + 2e \rightleftharpoons 2SO_4^{2-}$ | 2.01 | |
| $Co^{3+} + e \rightleftharpoons Co^{2+}$ | 1.82 | |
| $H_2O_2 + 2H^+ + 2e \rightleftharpoons 2H_2O$ | 1.77 | |
| $MnO_4^- + 4H^+ + 3e \rightleftharpoons MnO_2 + 2H_2O$ | 1.695 | |
| $Ce^{3+} + e \rightleftharpoons Ce^{3+}$ | | 1.70, 1 $F$ HClO$_4$; 1.61, 1 $F$ HNO$_3$; 1.44, 1 $F$ H$_2$SO$_4$ |
| $H_5IO_6 + H^+ + 2e \rightleftharpoons IO_3^- + 3H_2O$ | 1.6 | |
| $BrO_3^- + 6H^+ + 5e \rightleftharpoons \frac{1}{2}Br_2 + 3H_2O$ | 1.52 | |
| $MnO_4^- + 8H^+ + 5e \rightleftharpoons Mn^{2+} + 4H_2O$ | 1.51 | |
| $Mn^{3+} + e \rightleftharpoons Mn^{2+}$ | 1.51 | |
| $PbO_2 + 4H^+ + 2e \rightleftharpoons Pb^{2+} + 2H_2O$ | 1.455 | |
| $Cl_2 + 2e \rightleftharpoons 2Cl^-$ | 1.359 | |
| $Cr_2O_7^{2-} + 14H^+ + 6e \rightleftharpoons 2Cr^{3+} + 7H_2O$ | 1.33 | |
| $Tl^{3+} + 2e \rightleftharpoons Tl^+$ | 1.25 | 0.77, 1 $F$ HCl |
| $MnO_2 + 4H^+ + 2e \rightleftharpoons Mn^{2+} + 2H_2O$ | 1.23 | 1.24, 1 $F$ HClO$_4$ |
| $O_2 + 4H^+ + 4e \rightleftharpoons 2H_2O$ | 1.229 | |
| $IO_3^- + 6H^+ + 5e \rightleftharpoons \frac{1}{2}I_2 + 3H_2O$ | 1.195 | |
| $Br_2 + 2e \rightleftharpoons 2Br^-$ | 1 065 | 1 05, 4 $F$ HCl |
| $ICl_2^- + e \rightleftharpoons \frac{1}{2}I_2 + 2Cl^-$ | 1 06 | |
| $V(OH)_4^+ + 2H^+ + e \rightleftharpoons VO^{2+} + 3H_2O$ | 1 00 | 1 02, 1 $F$ HCl, HClO$_4$ |
| $HNO_2 + H^+ + e \rightleftharpoons NO + H_2O$ | 1.00 | |
| $NO_3^- + 3H^+ + 2e \rightleftharpoons HNO_2 + H_2O$ | 0.94 | 0.92, 1 $F$ HNO$_3$ |
| $2Hg^{2+} + 2e \rightleftharpoons Hg_2^{2+}$ | 0.920 | 0.907, 1 $F$ HClO$_4$ |
| $Cu^{2+} + I^- + e \rightleftharpoons CuI$ | 0.86 | |
| $Ag^+ + e \rightleftharpoons Ag$ | 0.799 | 0.228, 1 $F$ HCl; 0.792, 1 $F$ HClO$_4$; 0.77, 1$F$ H$_2$SO$_4$ |
| $Hg_2^{2+} + 2e \rightleftharpoons 2Hg$ | 0.789 | 0.274, 1 $F$ HCl; 0.776, 1 $F$ HClO$_4$; 0.674, 1 $F$ H$_2$SO$_4$ |
| $Fe^{3+} + e \rightleftharpoons Fe^{2+}$ | 0.771 | 0.700, 1 $F$ HCl; 0.732, 1 $F$ HClO$_4$; 0.68, 1 $F$ H$_2$SO$_4$ |
| $PtCl_4^{2-} + 2e \rightleftharpoons Pt + 4Cl^-$ | 0.73 | |
| $C_6H_4O_2$ (quinone) $+ 2H^+ + 2e \rightleftharpoons C_6H_4(OH)_2$ | 0.699 | 0.696, 1 $F$ HCl, H$_2$SO$_4$, HClO$_4$ |
| $O_2 + 2H^+ + 2e \rightleftharpoons H_2O_2$ | 0.682 | |
| $PtCl_6^{2-} + 2e \rightleftharpoons PtCl_4^{2-} + 2Cl^-$ | 0.68 | |

**TABLE 3.4.** *(Continued)*

| HALF REACTION | $E°$, VOLTS | FORMAL POTENTIAL, VOLTS |
|---|---|---|
| $MnO_4^- + e \rightleftharpoons MnO_4^{2-}$ | 0.564 | |
| $H_3AsO_4 + 2H^+ + 2e \rightleftharpoons H_3AsO_3 + H_2O$ | 0.559 | 0.577, 1 $F$ HCl, HClO$_4$ |
| $I_3^- + 2e \rightleftharpoons 3I^-$ | 0.536 | |
| $I_2 + 2e \rightleftharpoons 2I^-$ | 0.5355 | |
| $Cu^+ + e \rightleftharpoons Cu$ | 0.521 | |
| $H_2SO_3 + 4H^+ + 4e \rightleftharpoons S + 3H_2O$ | 0.45 | |
| $Ag_2CrO_4 + 2e \rightleftharpoons 2Ag + CrO^{2-}$ | 0.446 | |
| $VO^{2+} + 2H^+ + e \rightleftharpoons V^{3+} + H_2O$ | 0.361 | |
| $Fe(CN)_6^{3+} + e \rightleftharpoons Fe(CN)_6^{4-}$ | 0.36 | 0.71, 1 $F$ HCl; 0.72, 1 $F$ HClO$_4$, H$_2$SO$_4$ |
| $Cu^{2+} + 2e \rightleftharpoons Cu$ | 0.337 | |
| $UO_2^{2+} + 4H^+ + 2e \rightleftharpoons U^{3+} + 2H_2O$ | 0.334 | |
| $BiO^+ + 2H^+ + 3e \rightleftharpoons Bi + H_2O$ | $+$ 0.32 | |
| $Hg_2Cl_2 + 2e \rightleftharpoons 2Hg + 2Cl^-$ | 0.268 | 0.242, sat'd. KCl; 0.282, 1 $F$ KCl |
| $AgCl + e \rightleftharpoons Ag + Cl^-$ | 0.222 | 0.228, 1 $F$ KCl |
| $SO_4^{2-} + 4H^+ + 2e \rightleftharpoons H_2SO_3 + H_2O$ | 0.17 | |
| $Cu^{2+} + e \rightleftharpoons Cu^+$ | 0.153 | |
| $Sn^{3+} + 2e \rightleftharpoons Sn^{2+}$ | 0.15 | 0.14, 1 $F$ HCl |
| $S + 2H^+ + 2e \rightleftharpoons H_2S$ | 0.141 | |
| $TiO^{2+} + 2H^+ + e \rightleftharpoons Ti^{3+} + H_2O$ | 0.1 | 0 04, 1 $F$ H$_2$O$_4$ |
| $AgBr + e \rightleftharpoons Ag + Br^-$ | 0.095 | |
| $S_4O_6^{2-} + 2e \rightleftharpoons 2S_2O^{2-}$ | 0.08 | |
| $Ag(S_2O_3)^{3-} + e \rightleftharpoons Ag + 2S_2O_3^{2)}$ | 0.01 | |
| $2H^+ + 2e \rightleftharpoons H_2$ | 0.000 | $-$ 0.005, 1 $F$ HCl, HClO$_4$ |
| $Pb^{2+} + 2e \rightleftharpoons Pb$ | $-$ 0.126 | $-$ 0.14, 1 $F$ HClO$_4$; $-$ 0.29, 1 $F$ H$_2$SO$_4$ |
| $Sn^{2+} + 2e \rightleftharpoons Sn$ | $-$ 0.136 | $-$ 0.16, 1 $F$ HC.lO$_4$ |
| $AgI + e \rightleftharpoons Ag + I^-$ | $-$ 0.151 | |
| $CuI + e \rightleftharpoons Cu + I^-$ | $-$ 0.185 | |
| $N_2 + 5H^+ + 4e \rightleftharpoons N_2H_5^+$ | $-$ 0.23 | |
| $Ni^{2+} + 2e \rightleftharpoons Ni$ | $-$ 0.250 | |
| $V^{3+} + e \rightleftharpoons V^{2+}$ | $-$ 0.255 | $-$ 0.21, 1 $F$ HClO$_4$ |
| $Co^{2+} + 2e \rightleftharpoons Co$ | $-$ 0.277 | |
| $Ag(CN)_2^- + e \rightleftharpoons Ag + 2CN^-$ | $-$ 0.31 | |
| $Tl^+ + e \rightleftharpoons Tl$ | $-$ 0.336 | $-$ 0.551, 1 $F$ HCl; $-$ 0.33, 1 $F$ HClO$_4$, H$_2$SO$_4$ |
| $Ti^{3+} + e \rightleftharpoons Ti^{2+}$ | $-$ 0.37 | |
| $Cd^{2+} + 2e \rightleftharpoons Cd$ | $-$ 0.403 | |
| $Cr^{3+} + e \rightleftharpoons Cr^{2+}$ | $-$ 0.41 | |
| $Fe^{2+} + 2e \rightleftharpoons Fe$ | $-$ 0.440 | |
| $2CO_2(g) + 2H^+ + 2e \rightleftharpoons H_2C_2O_4$ | $-$ 0.49 | |
| $Cr^{3+} + 3e \rightleftharpoons Cr$ | $-$ 0.74 | |
| $Zn^{2+} + 2e \rightleftharpoons Zn$ | $-$ 0.763 | |
| $Mn^{2+} + 2e \rightleftharpoons Mn$ | $-$ 1.18 | |
| $Al^{3+} + 3e \rightleftharpoons Al$ | $-$ 1.66 | |
| $Mg^{2+} + 2e \rightleftharpoons Mg$ | $-$ 2.37 | |
| $Na^+ + e \rightleftharpoons Na$ | $-$ 2.71 | |
| $Ca^{2+} + 2e \rightleftharpoons Ca$ | $-$ 2.87 | |
| $Ba^{2+} + 2e \rightleftharpoons Ba$ | $-$ 2.90 | |
| $K^+ + e \rightleftharpoons K$ | $-$ 2.92 | |
| $Li^+ + e \rightleftharpoons Li$ | $-$ 3.04 | |

[1]The majority of $E°$ values are taken from Wendell M. Latimer, *Oxidation Potentials*, 2d ed., Englewood Cliffs, N.J.: Prentice Hall ,Inc., 1952. The formal potentials are from Ernest H. Swift, *Introductory Quantitative Analysis*. Englewood Cliffs, N.J.: Prentice Hall, Inc., 1950. With permission.

# Numerical Answers and Comments to Selected Problems

## Chapter 2

2.1. The reference level for the proton balance should be monohydrogen tartrate ion ($HT^-$) and water. Proton balance: $[H_2T] + [H_3O^+] + [T^{2-}] + [OH^-]$. Charge balance: $[H_3O^+] + [Na^+] = [HT^-] + 2[T^{2-}] + [OH^-]$. Mass balance: $[Na^+] = [H_2T] + [HT^-] + [T^{2-}]$.

2.2. A. $[H_3O^+] = [OH^-] + [OCl^-]$.
  C. $[H_3N(CH_2)_2NH_2^+] + 2[H_3N(CH_2)_2NH_3^{++}] = [HSO_4^-] + [H_2SO_4]$.
  D. $[NH_3] + [CH_3OH_2^+] = [CH_3O^-]$.

2.3. A. Eq. 2-46*, pH = 3.37
  B. Eq. 2-12; $[H_3O^+] = 1.01 \times 10^{-6}$, pH slightly less than 6.
  C,D. Use Fig. 2-10 as a guide: the acetic acid is essentially completely dissociated at this dilution (pH = 6); for $10^{-4}$ $F$ phenol, Eq. 2-48* should be used because the autoprotolysis of water is significant. pH = 6.85.
  E. A diprotic weak base. Solution as a monoprotic, very weak, dilute base using the basic analog of Eq. 2-48* gives pH = 7.28.
  F. Similar to E, except that the base is sufficiently strong that the basic analog of Eq. 2-46* is accurate. pH = 10.48.
  G. Amphiprotic. From the approximate equation pH = $\frac{1}{2}(pQ_1 + pQ_2)$, pH = 8.34.
  H. Monoprotic weak base. pH 11.13
  I,J. Both $NH_3 - NH_4^+$ buffers. (I): pH 9.94; (J): pH 9.43.
  K. An $H_3PO_4 - H_2PO_4^-$ buffer. pH 2.47

L. The product of the reaction is $1 \times 10^{-4}$ $F$ NaHPhthalate, an amphiprotic substance. From Eq. 2-54*, pH = 4.72. (Compare this result to that obtained using the approximate equation pH = $\frac{1}{2}(pK_1 + pK_2)$.)

2.4. The proton balance

$$[HOAc] + [H_3O^+] = [NH_3] + [OH^-]$$

simplifies to

$$[HOAc] = [NH_3]$$

On a logarithmic graph, this is true at pH 7.1. Formally, this system is reminiscent of an amphiprotic substance. Why?

2.5. $[HMNA^+] = (2.97 - 1.76) \times 10^{-3} = 1.21 \times 10^{-3}$ $M$.
From the values of $[HMNA^+]$, $[MNA]$, and $Q_a$ of $HMNA^+$,
$[H_3O^+] = 1.38 \times 10^{-3}$ $M$.
$[HClA] = 1.0 - 1.38 \times 10^{-3} = 0.9984$ $M$.
$[ClA^-] = 0.5 + 1.38 \times 10^{-3} = 0.5014$ $M$.
$Q_a$ of HClA = $2.75 \times 10^{-3}$
$pQ_a = 2.560$.

2.6. The factor $D$ in Eq. 2-58 is $1.38 \times 10^{-3}$, which is 0.138% of $C_A$, and 0.276% of $C_B$. Total error 0.41%.

2.7. The $pQ_a$ of this indicator is 5.5. Beginning of yellow tinge: pH = 5.5 − log 5 = 4.8. Last red tinge: pH = 5.5 + log 8 = 6.4.

2.8. pH = 6.11

$$
\begin{array}{ccc}
\text{H} & + & \text{H} \quad \pm \\
\end{array}
$$
2.9. $HOOCCH_2C-COOH$ ; $HOOCCH_2C-COO$ ,
$\quad\quad NH_3 \quad\quad\quad\quad\quad\quad NH_3$

$$\text{H} \quad\quad \pm$$
or $OOCCH_2C-COOH$
$\quad NH_3$

(Which do you think is more reasonable for the structure of this zwitterion?

$$\text{H} \quad\quad -$$
Note the location of the quadrivalent nitrogen.); $OOCCH_2C-COO$ ; and
$\quad\quad\quad\quad\quad\quad\quad\quad\quad\quad NH_3$

$$\text{H} \quad\quad 2-$$
$OOCCH_2C-COO$ .
$\quad NH_2$

2.10.  In the pH range between 5 and 9, glycine is predominately in the zwitterion form, but aspartic acid is an anion.

2.11.  Zwitterions in both cases; at the isoelectric point, the fraction present as zwitterion is maximum, i.e., at pH = $\frac{1}{2}(pK_1 + pK_2)$. Alanine: 6.11; aspartic acid, 3.0.

# Chapter 3

3.1.  | *Acid* | *Base* |
|---|---|
| $BF_3$ | $(CH_3)_3N$ |
| $Cu^{++}$ | $Cl^-$ |
| $H^+(proton)$ | $NH_3$ |
| $H^+(proton)$ | $OH^-$ |
| $Ag^+$ | $S_2O_3^{--}$ |
| $Co^{3+}$ | 2 en, $Cl^-$, $H_2O$ |

3.3.  Include a line for $[Cl^-]$ on your graph. The chloride balance

$$[Cl^-] + [HgCl_3^-] + 2[HgCl_4^{--}] = [HgCl^+] + 2[Hg^{++}]$$

simplifies, by inspection, to

$$[HgCl_3^-] = [HgCl^+]$$

at which point $HgCl_2$ is the predominant form, and $HgCl^+$ and $HgCl_3^-$ are each about $2 \times 10^{-3}$ $M$ (for a total of 0.4% of the mercury in solution).

3.4.  In this solution, $[SCN^-] = 1 \times 10^{-2}$ $M$, and the predominant species are $FeSCN^{+2}$ and $Fe(SCN)_2^+$.

3.5.  A. Above pH 9.7, where $NTA^{3-}$ predominates.

3.6.  A. $\alpha_{EBT} = 10^{-3.5}$; $Q'_{MgEBT} = 10^{7.0} \times 10^{-3.5} = 10^{3.5}$.

   B. $C_{Mg} = 10^{-3.5}$.

# Chapter 4

4.1.  A. $[Br_2]$

   B. $[Cl^-]$ (assuming that no species such as $Ag_2Cl^+$ form).

   C. $\frac{1}{2}[Ag^+]$ or $[CrO_4^{--}]$.

   D. $[CrO_4^{--}]$

   E. $[Br^-]$

   F. $[C_2O_4^{--}]$

   G. $[C_2O^{--}] + [HC_2O_4^-] + [H_2C_2O_4]$.

4.2.  $S = [Tl^+] = [I^-] = Q_{sp}^{1/2}$;

   $Q_{sp} = 6.3 \times 10^{-8}$

4.3.  $AgCl_2^-$; $S = 3 \times 10^{-6}$ $M$.

4.4.  The $Br^-$ balance should be

$$[Ag^+] = [Br^-] + [AgBr_2^-] + 2[AgBr_3^{--}] + 3[AgBr_4^{3-}]$$

and simplifies to

$$[Ag^+] = [Br^-]$$

the highest intersection. At that point, the predominant form of silver is $Ag^+$, $10^{-6.17}$ $M$ ($6.8 \times 10^{-7}$ $M$), and $[AgBr] = 10^{-7.96}$ $M$ ($1.1 \times 10^{-8}$ $M$), which is 1.6% of the total solubility. The rest contribute negligibly.

4.5.  A.  $S = 3.7 \times 10^{-3}$ $M$. The basicity of $HCOO^-$ is very weak, so that not enough protonates in pure water to affect the solubility measurably. (The reader should verify this by a rapid calculation of the pH of the solution, using the above solubility for $C_{\text{formate}}$.)

B.  $S = [Pb^{++}] = Q_{sp}/[HCOO^-]^2$.
$[HCOO^-] = 0.1 + 2S = ca.$ $0.1$ $M$.
$S = 2 \times 10^{-5}$ $M$, very small compared to 0.1.
However, in 0.01 $F$ $NaHCOO$, the same type of calculation yields

$$[HCOO^-] = ca.\ 0.01\ M, \text{ and } S = 2 \times 10^{-3}$$

which is not negligible with respect to 0.01. From Eq. 4-16 or, much simpler, by successive approximations, $S = 1.3 \times 10^{-3}$ $M$.

C.  $S = [Pb^{++}]$;

$$[HCOO^-] + [HCOOH] = 2S$$

$$[HCOO^-] = 2S \cdot \alpha_{HCOO}$$

At pH 4.0,

$$\alpha_{HCOO} = 0.64$$

$$Q_{sp} = S \cdot (2S\alpha)^2 = 4S^3\alpha^2$$

$$S = (Q_{sp}/4\alpha^2)^{1/3} = 5.0 \times 10^{-3}\ M$$

D.  Same approach as part C. The solubility in 0.01 $F$ $HNO_3$ would be less, since at equilibrium $[H_3O^+]$ would be less than 0.01 $M$.

4.6.  When $[OH^-]$ is greater than or equal to $(Q_{sp}/[Ca^{++}])^{1/2}$.

$$[OH^-] = 10^{-2}$$

$$pH = 12$$

4.9.  From Eq. 4-13, $S = 3 \times 10^{-7}$ $M$. This calculation ignores possible complexation of $Sr^{2+}$ by $PO_4^{3-}$, and the Brønsted basicity of $PO_4^{3-}$.

# Chapter 5

5.1.  B. Yes, dimerization in the nonpolar phase should increase the value of $E^{o/w}$ with increasing total formic acid.

E.  A plot of $E^{o/w}$ *vs.* $[HCOOH]_w$ should be linear, with slope $2Q_d(P^{o/w})^2$ and intercept $P^{o/w}$. You should expect $P^{o/w}$ to be quite small, and it is.

F.  Yes, qualitatively. The fraction dissociated decreases (Chapter 2), and thus $E^{o/w}$ increases, with increasing total formic acid present.

G.  A small amount of a strong acid was present in the water phase.

$$\text{H} \qquad\qquad\qquad\qquad \text{H}$$

5.2.  Let $HAl^{\pm}$ stand for $H_3CCCOO^-$, and $AlH$ stand for $H_3CCCOOH$.

$$\text{NH}_3 \qquad\qquad\qquad\qquad \text{NH}_2$$
$$+$$

Then $P^{w/o} = \dfrac{[HAl^{\pm}]_w}{[AlH]_o}$

and

$$E^{w/o} = \frac{[H_2Al^+]_w + [HAl^{\pm}]_w + [Al^-]_w}{[AlH]_o}$$

5.3.  The separation can be made when log $E^{w/o}$ for one base is positive ($E^{w/o}$ large) and for the other negative ($E^{w/o}$ small), which will be true at some pH below $pQ_a$ of the conjugate acid of the weaker base. This is seen most easily if you construct the suggested graph, putting both curves on the same graph.

5.4.  A. When $E^{m/s}$ is much greater than 1, log $R_F$ approaches 0. When $E^{m/s}$ is much less than 1, log $R_F$ approaches $E^{m/s}$.

B. $E^{m/s} = E^{o/w} = \dfrac{[B]_o}{[B]_w + [HB]_w} = \dfrac{[B]_o}{C_w} = \dfrac{[B]_o}{[B]_w/\alpha_B}$

$= P^{o/w} \cdot \alpha_B$

If log $R_F$ = log $E^{m/s}$, then
   log $R_F$ = log $P^{o/w}$ + log $\alpha_B$,
and the graph is a log $\alpha_B$ *vs.* pH graph (Cf. Figs. 2-5 and 5-2).

5.5.  They emerge in order of decreasing $R_F$.

*Note:* In 5.6 and 5.7, (+) indicates positive correlation and (−) indicates negative correlation.

5.6.  A. (−)
      B. (+)
      C. (−)
      D. (−)

5.7.  A. The larger $R_F$, the smaller the retention time.
  B.  (1) (+)
      (2) (+)
      (3) (−)
      (4) (−)

## Chapter 6

6.1.

| REACTION | $E^0$ | log $K$ |
|---|---|---|
| A.  $Fe^{++} + Ag^+ \rightarrow Fe^{3+} + Ag$ | +0.028 | 0.473 |
| B.  $2Li + F_2 \rightarrow 2Li^+ + 2F^-$ | +6.10 | 206 |
| *or* $Li + \frac{1}{2}F_2 \rightarrow Li^+ + F^-$ | +6.10 | 103 |
| C.  $Hg + Zn(OH)_2 \rightarrow HgO + Zn + H_2O$ | −1.37 | (See 6.3) |
| D.  $Cd + 2H^+ + 2OH^- \rightarrow Cd(OH)_2 + H_2$ | +0.829 | 28.0 |
| E.  $C_6H_4(OH)_2 + I_2 \rightarrow C_6H_4O_2 + 2H^+ + 2I^-$ | −0.163 | −5.51 |
| F.  $Pb + 2SO_4^{--} + PbO_2 + 4H^+ \rightarrow 2PbSO_4 + 2H_2O$ | +1.02 | 69.1 |

*Notes:* (1) In reactions C and D, the $E^0$ for the metal-metal hydroxide couples are obtained from the $E^0$ for the metal-metal ion couple and the $K_{sp}$ of the metal hydroxide, as in Eq. 6-44. (2) $E^0_{Hg^{2+},Hg}$ is not in Appendix 3, but may be calculated as follows:

| Couple | $E^0$ | $nE^0$ |
|---|---|---|
| $2Hg^{++} + 2e^- \rightarrow Hg_2^{++}$ | 0.920 | +1.840 |
| $Hg_2^{++} + 2e^- \rightarrow 2Hg$ | 0.789 | +1.578 |
| $2Hg^{++} + 4e^- \rightarrow 2Hg$ | - - - - | +3.418 |

$E^0_{Hg^{++},Hg} = nE^0$ in the last line divided by $n(4)$
$= +0.857$ V.

The rationale of this process is that when two equations are added, any *extensive* property associated with those reactions may also be added. $E$'s are intensive properties (Chapter 1) and are not additive, but the product $nE$ is proportional to electrical energy (recall that $nFE$ is the electrical work the cell can do) and is thus extensive and additive. $E^0$ of the sum reaction is recovered from the sum of $nE^0$'s by dividing by the number of electrons in the sum reaction. (3) Reaction C cannot reach equilibrium, so that the calculated value of $K$ is not useful. The situation is discussed further in the answer to problem 6.3. (4) Reactions with $E^0 > 0$ are spontaneous.

6.2.  Cd | Cd$^{++}$ || Pb$^{++}$ | Pb (+)
      (+) Ag | Ag$^+$ || Sn$^{4+}$, Sn$^{++}$ | Pt
      Ag | AgBr,Br$^-$ || Fe$^{3+}$,Fe$^{++}$ | Pt (+)
      Ag | AgBr,Br$^-$ || Ag$^+$ | Ag (+)


6.3.  A. Because $J$ approaches $K$.
      B. Zn  Zn(OH)$_2$,OH$^-$,HgO  Hg
         Zn + HgO + H$_2$O → Zn(OH)$_2$ + Hg

$$K = \left( \frac{(Zn)(OH_2)(Hg)}{(Zn)(HgO)(H_2O)} \right)_{eq.}$$

but as the cell is written, all components except H$_2$O have constant activity. Since H$_2$O is consumed in the cell reaction, the concentration of the hydroxide electrolyte reaches saturation, and then (H$_2$O) is also constant on further discharge of the cell. Thus, the proper quotient of activities $J$ can never reach the value of $K$, so that the reaction cannot reach equilibrium. The cell voltage is constant (constant $J$) at 1.37 V (problem 6.1) until the component present in smallest molar quantity is exhausted, at which point the reaction must stop, and $E$ becomes zero.

6.4.  AgCl → Ag$^+$ + Cl$^-$

$$E^0 = -0.577 \text{ V}; \log K = -9.75 = \log K_{sp}.$$

6.5.  A. From the Nernst Equation,

$$E_{Cd^{++},Cd} = E^0_{Cd^{++},Cd} + \frac{0.059}{2} \log [Cd^{++}].$$

Thus the right-hand electrode (C$_2$) is positive.

      B. Cd → Cd$^{++}$(C$_1$) + 2e$^-$
         2e$^-$ + Cd$^{++}$(C$_2$) → Cd
         —————————————————————
         Cd$^{++}$(C$_2$) → Cd$^{++}$(C$_1$)

Which way would Cd$^{++}$ ions diffuse if these two solutions were brought in contact?
      C. A possible answer is

$$\text{Hg | Hg}_2\text{SO}_4, \text{SO}_4^-(C_2) \text{ || SO}_4^-(C_1), \text{Hg}_2\text{SO}_4 \text{ | Hg.}$$

6.6.  The cell of 6.4 *is* a silver-ion concentration cell; [Ag$^+$] is governed in the right-hand half-cell by the solubility equilibrium.

6.7.    The equilibria are

$$Hg = Hg^{++} = HgCl_2$$

and

$$E = E^{0\prime}_{HgCl_2,Hg} = E^0_{HgCl,Hg} = E^0_{Hg^{++},Hg} + \frac{0.059}{2} \log (Hg^{++})$$

$$= E^0_{Hg^{++},Hg} + \frac{0.059}{2} \log [Hg^{++}] + \frac{0.059}{2} \log f_{Hg^{++}}$$

$$= E^0_{Hg^{++},Hg} + \frac{0.059}{2} \log [Hg^{++}]$$

In 1 $F$ HgCl$_2$,

$$[Hg^{2+}] = 1 \times 10^{-5}\ M\ (\text{problem 3.2}), \text{ so}$$

$$E^{0\prime}_{HgCl_2,Hg} = +0.857 + \frac{0.059}{2} \log (10^{-5})$$

$$= +0.709 \text{ V (Compare the formal potential for Hg}^{++},\text{Hg in } 1\ F\ \text{HCl}$$
in Appendix 3.)

6.8.    For any such electrode, the half-reaction is

$$M(OH)_n + ne^- \rightarrow M + nOH^-$$

and

$$E = E^0 - \frac{0.059}{n} \log (OH^-)^n$$

$$= E^0 - 0.059 \log (OH^-)$$

$$= E^0 - 0.059 \log K_w + 0.059 \log (H_3O^+)$$

$$= E^0 - 0.059 \log K_w - 0.059\ \text{pH}.$$

Since $n$ cancels out in this development, its value is immaterial.

6.9.    In the Nernst Equation,

$$E = E^{0\prime}_{Hg^{++},Hg} + \frac{0.059}{2} \log [Hg^{++}]$$

substitute for [Hg$^{++}$] using $Q_{HgY}$ and then for [Y] using $Q_{MY}$.

6.10. A. Use Eq. 6-44. $K_{sp} = 10^{-18.8}$.

B. Equilibrium I is independent of pH: $E = +0.337$ V. Equilibrium II follows the equation developed in problem 6.8:

$$E = E^0 - 0.059 \log K_w - 0.059 \text{ pH}$$

$$E = -0.219 + (0.059)(14.00) - 0.059 \text{ pH}$$

$$= (0.610 - 0.059 \text{ pH}) \text{ V}$$

C. $\qquad$ Cu $\rightarrow$ Cu$^{++}$ + 2e$^-$

$\dfrac{2e^- + Cu(OH)_2 \rightarrow Cu + 2OH^-}{Cu(OH)_2 \rightarrow Cu^{2+} + 2OH^-}$

D. The reverse of that in C.

E. The pH at which the lines of your graph intersect;

$$+0.337 = +0.610 - 0.059 \text{ pH}$$

$$\text{pH} = 4.61$$

F. Your final graph should look like this:

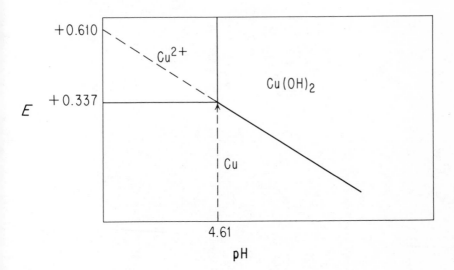

# APPENDIX 5

# Four-Place Common Logarithms

| N | 0 | 1 | 2 | 3 | 4 | 5 | 6 | 7 | 8 | 9 |
|----|------|------|------|------|------|------|------|------|------|------|
| 10 | 0000 | 0043 | 0086 | 0128 | 0170 | 0212 | 0253 | 0294 | 0334 | 0374 |
| 11 | 0414 | 0453 | 0492 | 0531 | 0569 | 0607 | 0645 | 0682 | 0719 | 0755 |
| 12 | 0792 | 0828 | 0864 | 0899 | 0934 | 0969 | 1004 | 1038 | 1072 | 1106 |
| 13 | 1139 | 1173 | 1206 | 1239 | 1271 | 1303 | 1335 | 1367 | 1399 | 1430 |
| 14 | 1461 | 1492 | 1523 | 1553 | 1584 | 1614 | 1644 | 1673 | 1703 | 1732 |
| 15 | 1761 | 1790 | 1818 | 1847 | 1875 | 1903 | 1931 | 1959 | 1987 | 2014 |
| 16 | 2041 | 2068 | 2095 | 2122 | 2148 | 2175 | 2201 | 2227 | 2253 | 2279 |
| 17 | 2304 | 2330 | 2355 | 2380 | 2405 | 2430 | 2455 | 2480 | 2504 | 2529 |
| 18 | 2553 | 2577 | 2601 | 2625 | 2648 | 2672 | 2695 | 2718 | 2742 | 2765 |
| 19 | 2788 | 2810 | 2833 | 2856 | 2878 | 2900 | 2923 | 2945 | 2967 | 2989 |
| 20 | 3010 | 3032 | 3054 | 3075 | 3096 | 3118 | 3139 | 3160 | 3181 | 3201 |
| 21 | 3222 | 3243 | 3263 | 3284 | 3304 | 3324 | 3345 | 3365 | 3385 | 3404 |
| 22 | 3424 | 3444 | 3464 | 3483 | 3502 | 3522 | 3541 | 3560 | 3579 | 3598 |
| 23 | 3617 | 3636 | 3655 | 3674 | 3692 | 3711 | 3729 | 3747 | 3766 | 3784 |
| 24 | 3802 | 3820 | 3838 | 3856 | 3874 | 3892 | 3909 | 3927 | 3945 | 3962 |
| 25 | 3979 | 3997 | 4014 | 4031 | 4048 | 4065 | 4082 | 4099 | 4116 | 4133 |
| 26 | 4150 | 4166 | 4183 | 4200 | 4216 | 4232 | 4249 | 4265 | 4281 | 4298 |
| 27 | 4314 | 4330 | 4346 | 4362 | 4378 | 4393 | 4409 | 4425 | 4440 | 4456 |
| 28 | 4472 | 4487 | 4502 | 4518 | 4533 | 4548 | 4564 | 4579 | 4594 | 4609 |
| 29 | 4624 | 4639 | 4654 | 4669 | 4683 | 4698 | 4713 | 4728 | 4742 | 4757 |
| 30 | 4771 | 4786 | 4800 | 4814 | 4829 | 4843 | 4857 | 4871 | 4886 | 4900 |
| 31 | 4914 | 4928 | 4942 | 4955 | 4969 | 4983 | 4997 | 5011 | 5024 | 5038 |
| 32 | 5051 | 5065 | 5079 | 5092 | 5105 | 5119 | 5132 | 5145 | 5159 | 5172 |
| 33 | 5185 | 5198 | 5211 | 5224 | 5237 | 5250 | 5263 | 5276 | 5289 | 5302 |
| 34 | 5315 | 5328 | 5340 | 5353 | 5366 | 5378 | 5391 | 5403 | 5416 | 5428 |
| 35 | 5441 | 5453 | 5465 | 5478 | 5490 | 5502 | 5514 | 5527 | 5539 | 5551 |
| 36 | 5563 | 5575 | 5587 | 5599 | 5611 | 5623 | 5635 | 5647 | 5658 | 5670 |
| 37 | 5682 | 5694 | 5705 | 5717 | 5729 | 5740 | 5752 | 5763 | 5775 | 5786 |
| 38 | 5798 | 5809 | 5821 | 5832 | 5843 | 5855 | 5866 | 5877 | 5888 | 5899 |
| 39 | 5911 | 5922 | 5933 | 5944 | 5955 | 5966 | 5977 | 5988 | 5999 | 6010 |
| 40 | 6021 | 6031 | 6042 | 6053 | 6064 | 6075 | 6085 | 6096 | 6107 | 6117 |
| 41 | 6128 | 6138 | 6149 | 6160 | 6170 | 6180 | 6191 | 6201 | 6212 | 6222 |
| 42 | 6232 | 6243 | 6253 | 6263 | 6274 | 6284 | 6294 | 6304 | 6314 | 6325 |
| 43 | 6335 | 6345 | 6355 | 6365 | 6375 | 6385 | 6395 | 6405 | 6415 | 6425 |
| 44 | 6435 | 6444 | 6454 | 6464 | 6474 | 6484 | 6493 | 6503 | 6513 | 6522 |
| 45 | 6532 | 6542 | 6551 | 6561 | 6571 | 6580 | 6590 | 6599 | 6609 | 6618 |
| 46 | 6628 | 6637 | 6646 | 6656 | 6665 | 6675 | 6684 | 6693 | 6702 | 6712 |
| 47 | 6721 | 6730 | 6739 | 6749 | 6758 | 6767 | 6776 | 6785 | 6794 | 6803 |
| 48 | 6812 | 6821 | 6830 | 6839 | 6848 | 6857 | 6866 | 6875 | 6884 | 6893 |
| 49 | 6902 | 6911 | 6920 | 6928 | 6937 | 6946 | 6955 | 6964 | 6972 | 6981 |
| 50 | 6990 | 6998 | 7007 | 7016 | 7024 | 7033 | 7042 | 7050 | 7059 | 7067 |
| 51 | 7076 | 7084 | 7093 | 7101 | 7110 | 7118 | 7126 | 7135 | 7143 | 7152 |
| 52 | 7160 | 7168 | 7177 | 7185 | 7193 | 7202 | 7210 | 7218 | 7226 | 7235 |
| 53 | 7243 | 7251 | 7259 | 7267 | 7275 | 7284 | 7292 | 7300 | 7308 | 7316 |
| 54 | 7324 | 7332 | 7340 | 7348 | 7356 | 7364 | 7372 | 7380 | 7388 | 7396 |
| N | 0 | 1 | 2 | 3 | 4 | 5 | 6 | 7 | 8 | 9 |

**10.0 — Four-Place Common Logarithms of Numbers — 54.9**

| N | 0 | 1 | 2 | 3 | 4 | 5 | 6 | 7 | 8 | 9 |
|---|---|---|---|---|---|---|---|---|---|---|
| 55 | 7404 | 7412 | 7419 | 7427 | 7435 | 7443 | 7451 | 7459 | 7466 | 7474 |
| 56 | 7482 | 7490 | 7497 | 7505 | 7513 | 7520 | 7528 | 7536 | 7543 | 7551 |
| 57 | 7559 | 7566 | 7574 | 7582 | 7589 | 7597 | 7604 | 7612 | 7619 | 7627 |
| 58 | 7634 | 7642 | 7649 | 7657 | 7664 | 7672 | 7679 | 7686 | 7694 | 7701 |
| 59 | 7709 | 7716 | 7723 | 7731 | 7738 | 7745 | 7752 | 7760 | 7767 | 7774 |
| 60 | 7782 | 7789 | 7796 | 7803 | 7810 | 7818 | 7825 | 7832 | 7839 | 7846 |
| 61 | 7853 | 7860 | 7868 | 7875 | 7882 | 7889 | 7896 | 7903 | 7910 | 7917 |
| 62 | 7924 | 7931 | 7938 | 7945 | 7952 | 7959 | 7966 | 7973 | 7980 | 7987 |
| 63 | 7993 | 8000 | 8007 | 8014 | 8021 | 8028 | 8035 | 8041 | 8048 | 8055 |
| 64 | 8062 | 8069 | 8075 | 8082 | 8089 | 8096 | 8102 | 8109 | 8116 | 8122 |
| 65 | 8129 | 8136 | 8142 | 8149 | 8156 | 8162 | 8169 | 8176 | 8182 | 8189 |
| 66 | 8195 | 8202 | 8209 | 8215 | 8222 | 8228 | 8235 | 8241 | 8248 | 8254 |
| 67 | 8261 | 8267 | 8274 | 8280 | 8287 | 8293 | 8299 | 8306 | 8312 | 8319 |
| 68 | 8325 | 8331 | 8338 | 8344 | 8351 | 8357 | 8363 | 8370 | 8376 | 8382 |
| 69 | 8388 | 8395 | 8401 | 8407 | 8414 | 8420 | 8426 | 8432 | 8439 | 8445 |
| 70 | 8451 | 8457 | 8463 | 8470 | 8476 | 8482 | 8488 | 8494 | 8500 | 8506 |
| 71 | 8513 | 8519 | 8525 | 8531 | 8537 | 8543 | 8549 | 8555 | 8561 | 8567 |
| 72 | 8573 | 8579 | 8585 | 8591 | 8597 | 8603 | 8609 | 8615 | 8621 | 8627 |
| 73 | 8633 | 8639 | 8645 | 8651 | 8657 | 8663 | 8669 | 8675 | 8681 | 8686 |
| 74 | 8692 | 8698 | 8704 | 8710 | 8716 | 8722 | 8727 | 8733 | 8739 | 8745 |
| 75 | 8751 | 8756 | 8762 | 8768 | 8774 | 8779 | 8785 | 8791 | 8797 | 8802 |
| 76 | 8808 | 8814 | 8820 | 8825 | 8831 | 8837 | 8842 | 8848 | 8854 | 8859 |
| 77 | 8865 | 8871 | 8876 | 8882 | 8887 | 8893 | 8899 | 8904 | 8910 | 8915 |
| 78 | 8921 | 8927 | 8932 | 8938 | 8943 | 8949 | 8954 | 8960 | 8965 | 8971 |
| 79 | 8976 | 8982 | 8987 | 8993 | 8998 | 9004 | 9009 | 9015 | 9020 | 9025 |
| 80 | 9031 | 9036 | 9042 | 9047 | 9053 | 9058 | 9063 | 9069 | 9074 | 9079 |
| 81 | 9085 | 9090 | 9096 | 9101 | 9106 | 9112 | 9117 | 9122 | 9128 | 9133 |
| 82 | 9138 | 9143 | 9149 | 9154 | 9159 | 9165 | 9170 | 9175 | 9180 | 9186 |
| 83 | 9191 | 9196 | 9201 | 9206 | 9212 | 9217 | 9222 | 9227 | 9232 | 9238 |
| 84 | 9243 | 9248 | 9253 | 9258 | 9263 | 9269 | 9274 | 9279 | 9284 | 9289 |
| 85 | 9294 | 9299 | 9304 | 9309 | 9315 | 9320 | 9325 | 9330 | 9335 | 9340 |
| 86 | 9345 | 9350 | 9355 | 9360 | 9365 | 9370 | 9375 | 9380 | 9385 | 9390 |
| 87 | 9395 | 9400 | 9405 | 9410 | 9415 | 9420 | 9425 | 9430 | 9435 | 9440 |
| 88 | 9445 | 9450 | 9455 | 9460 | 9465 | 9469 | 9474 | 9479 | 9484 | 9489 |
| 89 | 9494 | 9499 | 9504 | 9509 | 9513 | 9518 | 9523 | 9528 | 9533 | 9538 |
| 90 | 9542 | 9547 | 9552 | 9557 | 9562 | 9566 | 9571 | 9576 | 9581 | 9586 |
| 91 | 9590 | 9595 | 9600 | 9605 | 9609 | 9614 | 9619 | 9624 | 9628 | 9633 |
| 92 | 9638 | 9643 | 9647 | 9652 | 9657 | 9661 | 9666 | 9671 | 9675 | 9680 |
| 93 | 9685 | 9689 | 9694 | 9699 | 9703 | 9708 | 9713 | 9717 | 9722 | 9727 |
| 94 | 9731 | 9736 | 9741 | 9745 | 9750 | 9754 | 9759 | 9763 | 9768 | 9773 |
| 95 | 9777 | 9782 | 9786 | 9791 | 9795 | 9800 | 9805 | 9809 | 9814 | 9818 |
| 96 | 9823 | 9827 | 9832 | 9836 | 9841 | 9845 | 9850 | 9854 | 9859 | 9863 |
| 97 | 9868 | 9872 | 9877 | 9881 | 9886 | 9890 | 9894 | 9899 | 9903 | 9908 |
| 98 | 9912 | 9917 | 9921 | 9926 | 9930 | 9934 | 9939 | 9943 | 9948 | 9952 |
| 99 | 9956 | 9961 | 9965 | 9969 | 9974 | 9978 | 9983 | 9987 | 9991 | 9996 |
| N | 0 | 1 | 2 | 3 | 4 | 5 | 6 | 7 | 8 | 9 |

**55.0 — Four-Place Common Logarithms of Numbers — 99.9**

# Index